Holt Mathematics

Know-It Notebook™

HOLT, RINEHART AND WINSTON
A Harcourt Education Company
Orlando • **Austin** • New York • San Diego • London

Copyright © by Holt, Rinehart and Winston

All rights reserved. No part of this publication may be reproduced or transmitted in any form or by any means, electronic or mechanical, including photocopy, recording, or any information storage and retrieval system, without permission in writing from the publisher.

Requests for permission to make copies of any part of the work should be mailed to the following address: Permissions Department, Holt, Rinehart and Winston, 10801 N MoPac Expressway, Building 3, Austin, Texas 78759.

HOLT and the **"Owl Design"** are trademarks licensed to Holt, Rinehart and Winston, registered in the United States of America and/or other jurisdictions.

Printed in the United States of America

If you have received these materials as examination copies free of charge, Holt, Rinehart and Winston retains title to the materials and they may not be resold. Resale of examination copies is strictly prohibited and is illegal.

Possession of this publication in print format does not entitle users to convert this publication, or any portion of it, into electronic format.

ISBN 0-03-078327-5

12 13 14 15 16 1418 15 14 13 12 11 10
4500247730

Copyright © by Holt, Rinehart and Winston. All rights reserved.

Contents

Using the Know-It Notebook	v
Note Taking Strategies	vii
Lesson 1-1	1
Lesson 1-2	4
Lesson 1-3	7
Lesson 1-4	9
Lesson 1-5	11
Lesson 1-6	13
Lesson 1-7	16
Lesson 1-8	18
Lesson 1-9	20
Lesson 1-10	22
Lesson 1-11	25
Lesson 1-12	27
Ch 1 Chapter Review	30
Ch 1 Big Ideas	34
Lesson 2-1	35
Lesson 2-2	37
Lesson 2-3	40
Lesson 2-4	42
Lesson 2-5	44
Lesson 2-6	46
Lesson 2-7	49
Lesson 2-8	52
Lesson 2-9	55
Lesson 2-10	58
Lesson 2-11	60
Ch 2 Chapter Review	63
Ch 2 Big Ideas	66
Lesson 3-1	67
Lesson 3-2	69
Lesson 3-3	71
Lesson 3-4	73
Lesson 3-5	75
Lesson 3-6	78
Lesson 3-7	82
Lesson 3-8	84
Lesson 3-9	86
Lesson 3-10	88
Lesson 3-11	90
Lesson 3-12	93
Ch 3 Chapter Review	95
Ch 3 Big Ideas	99
Lesson 4-1	100
Lesson 4-2	102
Lesson 4-3	104
Lesson 4-4	107
Lesson 4-5	109
Lesson 4-6	111
Ch 4 Chapter Review	113
Ch 4 Big Ideas	116
Lesson 5-1	117
Lesson 5-2	119
Lesson 5-3	122
Lesson 5-4	125
Lesson 5-5	127
Lesson 5-6	129
Lesson 5-7	131
Lesson 5-8	133
Lesson 5-9	136
Ch 5 Chapter Review	139
Ch 5 Big Ideas	142
Lesson 6-1	143
Lesson 6-2	145
Lesson 6-3	147
Lesson 6-4	149
Lesson 6-5	151
Lesson 6-6	153
Lesson 6-7	156
Ch 6 Chapter Review	159
Ch 6 Big Ideas	161
Lesson 7-1	162
Lesson 7-2	165
Lesson 7-3	169
Lesson 7-4	172
Lesson 7-5	174
Lesson 7-6	176
Lesson 7-7	178
Lesson 7-8	180
Lesson 7-9	182
Lesson 7-10	184
Ch 7 Chapter Review	185
Ch 7 Big Ideas	188

Lesson 8-1	189	Lesson 11-1	250
Lesson 8-2	191	Lesson 11-2	252
Lesson 8-3	193	Lesson 11-3	253
Lesson 8-4	195	Lesson 11-4	256
Lesson 8-5	197	Lesson 11-5	258
Lesson 8-6	199	Lesson 11-6	260
Lesson 8-7	201	Lesson 11-7	262
Lesson 8-8	203	Ch 11 Chapter Review	264
Lesson 8-9	205	Ch 11 Big Ideas	267
Lesson 8-10	207	Lesson 12-1	268
Lesson 8-11	210	Lesson 12-2	270
Ch 8 Chapter Review	212	Lesson 12-3	273
Ch 8 Big Ideas	215	Lesson 12-4	275
Lesson 9-1	216	Lesson 12-5	277
Lesson 9-2	218	Lesson 12-6	279
Lesson 9-3	220	Lesson 12-7	281
Lesson 9-4	222	Ch 12 Chapter Review	283
Lesson 9-5	224	Ch 12 Big Ideas	285
Lesson 9-6	226		
Lesson 9-7	229		
Lesson 9-8	231		
Ch 9 Chapter Review	234		
Ch 9 Big Ideas	236		
Lesson 10-1	237		
Lesson 10-2	239		
Lesson 10-3	241		
Lesson 10-4	242		
Lesson 10-5	245		
Ch 10 Chapter Review	247		
Ch 10 Big Ideas	249		

USING THE *KNOW-IT NOTEBOOK*™

This *Know-It Notebook* will help you take notes, organize your thinking, and study for quizzes and tests. There are *Know-It Notes*™ pages for every lesson in your textbook. These notes will help you identify important mathematical information that you will need later.

Know-It Notes
Lesson Objectives
A good note-taking practice is to know the objective the content covers.
Vocabulary
Another good note-taking practice is to keep a list of the new vocabulary.
- Use the page references or the glossary in your textbook to find each definition.
- Write each definition on the lines provided.

Additional Examples
Your textbook includes examples for each math concept taught. Additional examples in the *Know-It Notebook* help you take notes so you remember how to solve different types of problems.
- Take notes as your teacher discusses each example.
- Write notes in the blank boxes to help you remember key concepts.
- Write final answers in the shaded boxes.

Try This
Complete the Try This problems that follow some lessons. Use these to make sure you understand the math concepts covered in the lesson.
- Write each answer in the space provided.
- Check your answers with your teacher or another student.
- Ask your teacher to help you understand any problem that you answered incorrectly.

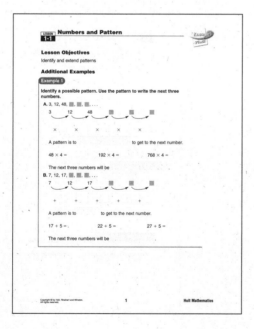
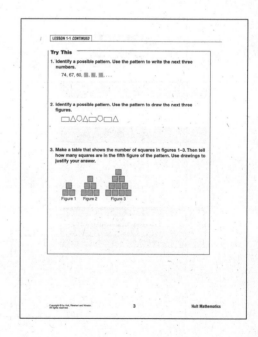

Chapter Review

Complete Chapter Review problems that follow each chapter. This is a good review before you take the chapter test.

- Write each answer in the space provided.
- Check your answers with your teacher or another student.
- Ask your teacher to help you understand any problem that you answered incorrectly.

Big Ideas

The Big Ideas have you summarize the important chapter concepts in your own words. You must think about and understand ideas to put them in your own words. This will also help you remember them.

- Write each answer in the space provided.
- Check your answers with your teacher or another student.
- Ask your teacher to help you understand any question that you answered incorrectly.

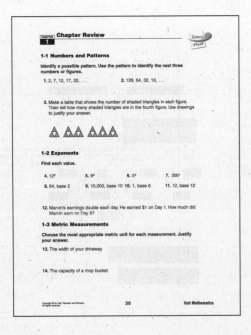

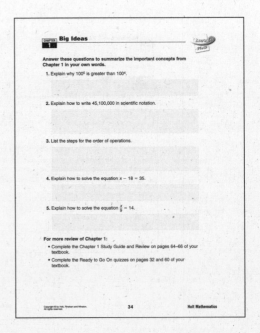

NOTE TAKING STRATEGIES

Taking good notes is very important in many of your classes and will be even more important when you take college classes. This notebook was designed to help you get started. Here are some other steps that can help you take good notes.

Getting Ready

1. Use a loose-leaf notebook. You can add pages to this where and when you want to. It will help keep you organized.

During the Lecture

2. If you are taking notes during a lecture, write the big ideas. Use abbreviations to save time. Do not worry about spelling or writing every word. Use headings to show changes in the topics discussed. Use numbering or bullets to organize supporting ideas under each topic heading. Leave space before each new heading so that you can fill in more information later.

After the Lecture

3. As soon as possible after the lecture, read through your notes and add any information that will help you understand them when you review later. You should also summarize the information into key words or key phrases. This will help your comprehension and will help you process the information. These key words and key phrases will be your memory cues when you are reviewing for or taking a test. At this time you may also want to write questions to help clarify the meaning of the ideas and facts.

4. Read your notes out loud. As you do this, state the ideas in your own words and do as much as you can by memory. This will help you remember and will also help with your thinking process. This activity will help you understand the information.

5. Reflect upon the information you have learned. Ask yourself how new information relates to information you already know. Ask how this relates to your personal experience. Ask how you can apply this information and why it is important.

Before the Test

6. Review your notes. Don't wait until the night before the test to review. Do frequent reviews. Don't just read through your notes. Put the information in your notes into your own words. If you do this you will be able to connect the new material with material you already know, and you will be better prepared for tests. You will have less test anxiety and better recall.

7. Summarize your notes. This should be in your own words and should only include the main points you need to remember. This will help you internalize the information.

LESSON 1-1 Numbers and Pattern

Lesson Objectives

Identify and extend patterns

Additional Examples

Example 1

Identify a possible pattern. Use the pattern to write the next three numbers.

A. 3, 12, 48, ▪, ▪, ▪, . . .

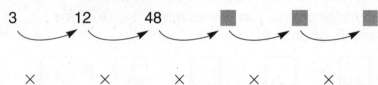

A pattern is to _____ to get to the next number.

48 × 4 = ____   192 × 4 = ____   768 × 4 = ____

The next three numbers will be _____ .

B. 7, 12, 17, ▪, ▪, ▪, . . .

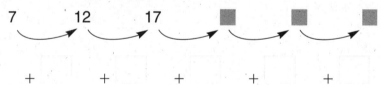

A pattern is to _____ to get to the next number.

17 + 5 = ____   22 + 5 = ____   27 + 5 = ____

The next three numbers will be _____ .

LESSON 1-1 CONTINUED

Example 2

Identify a possible pattern. Use the pattern to draw the next three figures.

The pattern is □△▽△□▽□△

The next three figures will be _____.

Example 3

Make a table that shows the number of triangles in figures 1–5 in Example 3. Then tell how many triangles are in the seventh figure of the pattern. Use drawings to justify your answer.

      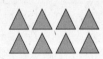 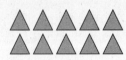

Figure 1   Figure 2   Figure 3   Figure 4   Figure 5

Make a table that shows the number of triangles in each figure.

Figure	1	2	3	4	5
Number of Triangles					

☐ + ☐ + ☐ + ☐

The pattern is to _____ triangles each time.

Figure 6 has 10 + 2 = _____ triangles.   Figure 7 has 12 + 2 = _____ triangles.

Figure 7 will have _____ triangles.

Holt Mathematics

LESSON 1-1 CONTINUED

Try This

1. Identify a possible pattern. Use the pattern to write the next three numbers.

   74, 67, 60, ■, ■, ■, . . .

2. Identify a possible pattern. Use the pattern to draw the next three figures.

   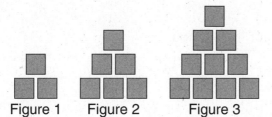

3. Make a table that shows the number of squares in figures 1–3. Then tell how many squares are in the fifth figure of the pattern. Use drawings to justify your answer.

   Figure 1   Figure 2   Figure 3

LESSON 1-2: Exponents

Lesson Objectives
Represent numbers by using exponents

Vocabulary

power (p. 10)

exponent (p. 10)

base (p. 10)

Additional Examples

Example 1

Find each value.

A. 4^4

$4^4 = 4 \cdot 4 \cdot 4 \cdot 4$ Use ☐ as a factor ☐ times.

= ☐

B. 7^3

$7^3 = 7 \cdot 7 \cdot 7$ Use ☐ as a factor ☐ times.

= ☐

C. 19^1

$19^1 = 19$ Use ☐ as a factor ☐ time.

= ☐

LESSON 1-2 CONTINUED

Example 2

Write each number using an exponent and the given base.

A. 625, base 5

625 = 5 · 5 · 5 · 5   is used as a factor ___ times.

= ___

B. 64, base 2

64 = 2 · 2 · 2 · 2 · 2 · 2   is used as a factor ___ times.

= ___

Example 3

On Monday, Erik tells 3 people a secret. The next day each of them tells 3 more people. If this pattern continues, how many people besides Erik will know the secret on Friday?

On Monday, ___ people know the secret.

On Tuesday, ___ times as many people know as those who knew on Monday.

On Wednesday, ___ times as many people know as those who knew on Tuesday.

On Thursday, ___ times as many people know as those who knew on Wednesday.

On Friday, 3 times as many people know as those who knew on ___.

Each day the number of people is ___ times greater.

3 · 3 · 3 · 3 · 3 = ___ = ___

On Friday, ___ people besides Erik will know the secret.

Holt Mathematics

LESSON 1-2 CONTINUED

Try This

1. Find the value.

    3^3

2. Write the number using an exponent and the given base.

    2,401, base 7

3. In a game, a contestant had a starting score of one point. She doubled her score every turn for four turns. Write her score after four turns as a power. Then find her score.

LESSON 1-3 Metric Measurement

Lesson Objectives

Identify, convert, and compare metric units

Additional Examples

Example 1

Choose the most appropriate metric unit for each measurement. Justify your answer.

A. The amount of water a runner drinks each day

_____ – the amount of water a runner drinks is similar to the amount of water in several large _____.

B. The length of a boat

_____ – the length of a boat is similar to the width of several _____.

C. The mass of a car

_____ – the mass of a car is similar to the mass of several _____.

Example 2

Convert each measure.

A. 530 cL to liters

530 cL = (530 ÷ _____) L   100 cL = 1 L, so divide by _____.

= _____ L   Move the decimal point _____ places to the _____.

LESSON 1-3 CONTINUED

Convert each measure.

B. 1,070 g to milligrams

1,070 g = (1,070 × ☐) mg   1 g = 1,000 mg, so multiply by ☐.

= ☐ mg   Move the decimal point ☐ places to the ☐.

Example 3

Elizabeth purchases one pumpkin that weighs 3 kg and another that weighs 2,150 g. Which pumpkin weighs more? Use estimation to explain why your answer makes sense.

Convert 2,150 g to kilograms.

2,150 g = (2,150 ÷ ☐) kg   1,000 g = 1 kg, so divide by ☐.

= ☐ g   Move the decimal point ☐ places to the ☐.

The ☐ pumpkin weighs more.

Check

2,150 g is about ☐ g or 2 kg   Round 2,150 to the nearest ☐.

3 kg > 2 kg, so the answer makes sense.

Try This

1. Choose the most appropriate metric unit for the measurement. Justify your answer.

    The amount of water in a kitchen sink

LESSON 1-4 Applying Exponents

Lesson Objectives

Multiply by powers of ten and express large numbers in scientific notation

Vocabulary

scientific notation (p. 18)

Additional Examples

Example 1

Multiply $14 \cdot 10^3$.

A. Evaluate the power.

$14 \cdot 10^3 = 14 \cdot ($ ___ $\cdot$ ___ $\cdot$ ___ $)$ Use ___ as a factor ___ times.

$= 14 \cdot$ ___   Multiply.

$=$ ___

B. Use mental math.

$14 \cdot 10^3 = 14.000$ Move the decimal point ___ places.

$=$ ___   Add ___ zeros.

Example 2

Write 4,340,000 in scientific notation.

$4{,}340{,}000 = 4{,}340{,}000$ Move the decimal point to get a number between ___ and ___.

$=$ ___   The exponent is equal to the number of places the decimal point is moved.

Example 3

The population of China in the year 2000 was estimated to be about 1.262×10^9. Write this number in standard form.

$1.262 \times 10^9 = 1.262000000$ Since the exponent is ____, move the decimal point ____ places to the ____.

= ____

The population of China was about ____ people.

Example 4

In 2005, the population of Mexico was $1.06 \cdot 10^8$ and the population of Brazil was $1.86 \cdot 10^8$. In which country do more people live?

To compare numbers written in scientific notation, first compare the ____. If the exponents are equal then compare the ____ portion of the number.

Mexico: $1.06 \cdot 10^8$ Brazil: $1.86 \cdot 10^8$

The exponents are the same, so compare the decimal portion of the number.

1.06 ____ 1.86, so $1.06 \cdot 10^8$ ____ $1.86 \cdot 10^8$.

More people live in ____.

Try This

1. Multiply $12 \cdot 10^2$.

2. Write 8,421,000 in scientific notation.

LESSON 1-5 Order of Operations

Lesson Objectives
Use the order of operations to simplify numerical expressions

Vocabulary
numerical expression (p. 23)

order of operations (p. 23)

Additional Examples

Example 1

Simplify each expression.

A. $3 + 15 \div 5$

   $3 + 15 \div 5$ Divide.

   $3 +$ ☐   Add.

   ☐

B. $44 - 14 \div 2 \cdot 4 + 6$

   $44 - 14 \div 2 \cdot 4 + 6$ _____ and _____ from left to right.

   $44 -$ ☐ $\cdot 4 + 6$

   $44 -$ ☐ $+ 6$ _____ and _____ from left to right.

   ☐ $+ 6$

   ☐

LESSON 1-5 CONTINUED

Example 2

Simplify each expression.

A. $42 - (3 \cdot 4) \div 6$

$42 - (3 \cdot 4) \div 6$ Perform the operation inside the _____.

$42 - \boxed{} \div 6$

$42 - \boxed{}$

$\boxed{}$

B. $[(26 - 4 \cdot 5) + 6]^2$

$[(26 - 4 \cdot 5) + 6]^2$ The parentheses are inside the brackets, so perform the operations inside the _____ first.

$[(26 - \boxed{}) + 6]^2$

$[\boxed{} + \boxed{}]^2$

$\boxed{}^2$

$\boxed{}$

Example 3

Sandy runs 4 miles per day. She ran 5 days during the first week of the month. She ran only 3 days each week for the next 3 weeks. Simplify the expression $(5 + 3 \cdot 3) \cdot 4$ to find how many miles she ran last month.

$(5 + 3 \cdot 3) \cdot 4$ Perform the operations inside the _____ first.

$(5 + \boxed{}) \cdot 4$ Add.

$\boxed{} \cdot 4$ Multiply.

$\boxed{}$

Sandy ran _____ miles last month.

LESSON 1-6 Properties

Lesson Objectives

Identify properties of rational numbers and use them to simplify numerical expressions

Vocabulary

Commutative Property (p. 28)

Associative Property (p. 28)

Identity Property (p. 28)

Distributive Property (p. 29)

Additional Examples

Example 1

Tell which property is represented.

A. (2 · 6) · 1 = 2 · (6 · 1)

  (2 · 6) · 1 = 2 · (6 · 1)   The numbers are _____.

B. 3 + 0 = 3

  3 + 0 = 3   The sum of 3 and ____ is 3.

C. 7 + 9 = 9 + 7

  7 + 9 = 9 + 7   The order of the numbers is _____.

LESSON 1-6 CONTINUED

Example 2

Simplify each expression. Justify each step.

A. 21 + 16 + 9

21 + 16 + 9 = ☐ + ☐ + 9   _____ Property

= 16 + (21 + 9)   _____ Property

= 16 + ☐   Add.

= ☐

B. 20 · 9 · 5

20 · 9 · 5 = 20 · ☐ · ☐   _____ Property

= (20 · 5) · 9   _____ Property

= ☐ · 9   Multiply.

= ☐

Example 3

Use the Distributive Property to find 6(54).

Method 1: 6(54) = 6(☐ + ☐)   Rewrite 54 as ☐ + ☐.

= (6 · ☐) + (6 · ☐)   Use the _____ Property.

= ☐ + ☐   Multiply.

= ☐   Add.

LESSON 1-6 CONTINUED

Use the Distributive Property to find 6(54).

Method 2: 6(54) = 6(☐ − ☐)   Rewrite 54 as ☐ − ☐.

= (6 · ☐) − (6 · ☐)   Use the _____ Property.

= ☐ − ☐   Multiply.

= ☐   Subtract.

Try This

1. Tell which property is represented.

    (5 + 6) + 3 = 5 + (6 + 3)

2. Simplify the expression. Justify each step.

    215 + 73 + 85

3. Use the Distributive Property to find 7(18).

15

Holt Mathematics

LESSON 1-7: Variables and Algebraic Expressions

Lesson Objectives

Evaluate algebraic expressions

Vocabulary

variable (p. 34)

constant (p. 34)

algebraic expression (p. 34)

evaluate (p. 34)

Additional Examples

Example 1

Evaluate $k + 9$ for each value of k.

A. $k = 5$ $k + 9$

☐ $+ 9$ Substitute ☐ for k.

☐   Add.

B. $k = 2$ $k + 9$

☐ $+ 9$ Substitute ☐ for k.

☐   Add.

LESSON 1-7 CONTINUED

Example 2

Evaluate each expression for the given value of the variable.

A. $4x - 3$ for $x = 2$

    $4() - 3$ Substitute ___ for x.

                  Multiply.

          -3 Subtract.

B. $s \div 5 + s$, for $s = 15$

    ___ $\div 5 +$ ___   Substitute 15 for ___.

                    Divide.

    ___ $+ 15$ Add.

Example 3

Evaluate $\frac{6}{a} + 4b$, **for** $a = 3$ **and** $b = 2$.

$\frac{6}{a} + 4b$

    $\frac{}{} + 4()$ Substitute ___ for a and ___ for b.

                   Divide and multiply from ___ to ___.

    ___ $+$ ___   Add.

Try This

1. Evaluate $a + 6$ for the value of a.

    $a = 3$

LESSON 1-8: Translate Words into Math

Lesson Objectives
Translate words into numbers, variables, and operations

Additional Examples

Example 1

Write each phrase as an algebraic expression.

A. the quotient of a number and 4

quotient means "☐"

☐

B. w increased by 5

increased by means "☐"

☐

C. the difference of 3 times a number and 7

☐ · ☐ − 7

☐

D. the quotient of 4 and a number, increased by 10

☐

LESSON 1-8 CONTINUED

Example 2

A. Mr. Campbell drives at 55 mi/h. Write an expression for how far he can drive in h hours.

You need to put _____ parts together. This involves multiplication.

55mi/h · h hours = _____

B. On a history test Maritza scored 50 points on the essay. Besides the essay, each short-answer question was worth 2 points. Write an expression for her total points if she answered q short-answer questions correctly.

The total points include _____ points for each short-answer question.

Multiply to put _____ parts together: _____

In addition to the points for short-answer questions, the total points included _____ points on the essay.

Add to put the parts together: _____

Try This

1. Write the phrase as an algebraic expression.

    4 times the difference of y and 8

2. Julie Ann works on an assembly line building computers. She can assemble 8 units an hour. Write an expression for the number of units she can produce in h hours.

LESSON 1-9: Simplifying Algebraic Expressions

Lesson Objectives
Simplify algebraic expressions

Vocabulary
term (p. 42) _____

coefficient (p. 42) _____

Additional Examples

Example 1
Identify like terms in the list.

$3t \quad 5w^2 \quad 7t \quad 9v \quad 4w^2 \quad 8v$

Look for like _____ with like _____.

$3t \quad 5w^2 \quad 7t \quad 9v \quad 4w^2 \quad 8v$

Like terms: _____

Example 2
Simplify. Justify your steps using the Commutative, Associative, and Distributive Properties when necessary.

A. $6t - 4t$ \qquad $6t$ and $4t$ are _____ terms.

$6t - 4t$ \qquad _____ the coefficients.

B. $45x - 37y + 87$

In this expression, there are _____ to combine.

LESSON 1-9 CONTINUED

Example 3

Write an expression for the perimeter of the triangle. Then simplify the expression.

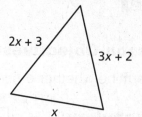

$x + 3x + 2 + 2x + 3$ Write an _____ using the side lengths.

(___ + ___ + ___) + (___ + ___)   Identify and _____ like terms.
_____   _____   like terms.

Try This

1. Identify like terms in the list.

   $2x \quad 4y^3 \quad 8x \quad 5z \quad 5y^3 \quad 8z$

2. Simplify. Justify your steps using the Commutative, Associative, and Distributive Properties when necessary.

   $4x^2 + 4y + 3x^2 - 4y + 2x^2 + 5$

3. Write an expression for the perimeter of the triangle. Then simplify the expression.

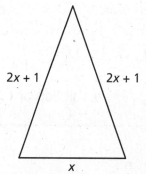

LESSON 1-10: Equations and Their Solutions

Lesson Objectives

Determine whether a number is a solution of an equation

Vocabulary

equation (p. 46) _____

solution (p. 46) _____

Additional Examples

Example 1

Determine whether the given value of the variable is a solution of $t + 9 = 17$.

A. 26

$t + 9 = 17$

☐ $+ 9 \stackrel{?}{=} 17$ Substitute ____ for t.

☐ $\stackrel{?}{=} 17$ ✗

26 _____ a solution of $t + 9 = 17$.

B. 8

$t + 9 = 17$

☐ $+ 9 \stackrel{?}{=} 17$ Substitute 8 for ____.

☐ $\stackrel{?}{=} 17$ ✓

8 _____ a solution of $t + 9 = 17$.

LESSON 1-10 CONTINUED

Example 2

Mrs. Jenkins had $32 when she returned home from grocery shopping. If she spent $17 at the supermarket, did she have $52 or $49 before she went shopping?

If x represents the amount of [   ] she had before she went shopping, then $x - 17 = 32$.

$52

$x - 17 = 32$

[   ] $- 17 \stackrel{?}{=} 32$ Substitute 52 for [   ].

[   ] $\stackrel{?}{=} 32$ ✗

$49

$x - 17 = 32$

[   ] $- 17 \stackrel{?}{=} 32$ Substitute [   ] for x.

[   ] $\stackrel{?}{=} 32$ ✓

Mrs. Jenkins had $[   ] before she went shopping.

Example 3

Which situation best matches the equation $5 + 2x = 13$?

Situation A:

Admission to the country fair costs $5 and rides cost $2 each. Mike spent a total of $13. How many rides did he go on?

$5 admission → [   ]

$2 for each ride → + [   ]

Mike spent $13 in all, so [   ] = 13. [   ] matches the equation.

LESSON 1-10 CONTINUED

Which situation best matches the equation $5 + 2x = 13$?

Situation B:

Admission to the county fair costs $2 and rides cost $5 each. Mike spent a total of $13. How many rides did he go on?

The variable x represents the number of rides Mike went on.

$5 per ride →

Since _____ is not a term in the given equation, _____ does not match the equation.

Try This

1. Determine whether the number is a solution of $x - 5 = 12$.

    22

2. During a scavenger hunt James found 34 items, 9 more than Billy. The equation $34 = e + 9$ can be used to represent the number of items James found. Did Billy find 43 items or 25 items?

3. Which situation best matches the equation $12h + 50 = 236$?

    Situation A: Darci earns $12 plus $50 an hour. She earned a total of $236 last week. How many hours did she work?

    Situation B: Darci earns $12 an hour plus $50. She earned a total of $236 last week. How many hours did she work?

Addition and Subtraction Equations

LESSON 1-11

Lesson Objectives

Solve one-step equations by using addition or subtraction

Vocabulary

Addition Property of Equality (p. 52)

inverse operations (p. 52)

Subtraction Property of Equality (p. 53)

Additional Examples

Example 1

Solve the equation $b - 7 = 24$. Check your answer.

$b - 7 = 24$ Think: 7 is _____ from b, so
$\underline{+7 \ +7}$ _____ 7 to both sides to _____ b.
$b =$

Check

$b - 7 = 24$

$\underline{} - 7 \stackrel{?}{=} 24$ Substitute _____ for b.

$ \stackrel{?}{=} 24$ ✓   _____ is a solution.

LESSON 1-11 CONTINUED

Example 2

Solve the equation $t + 14 = 29$.

$t + 14 = 29$
$-14 \quad -14$
$t = $

Think: ☐ is added to t, so subtract ☐ from both sides to isolate t.

Example 3

The Giants scored 13 points in a game against Dallas. They scored 7 points for a touchdown and the rest of their points for field goals. How many points did they score on field goals?

Let f represent the field goal points.

7 points + field goal points = points scored

$7 \quad + \quad f \quad = \quad$ ☐

$7 + f = $ ☐

$-7 -7$ Subtract ☐ from both sides to ☐ f.
$f = 6$

They scored ☐ points on field goals.

Try This

1. Solve the equation $y - 3 = 21$. Check your answer.

2. Solve the equation $x + 11 = 36$. Check your answer.

LESSON 1-12: Multiplication and Division Equations

Lesson Objectives

Solve one-step equations by using multiplication or division

Vocabulary

Multiplication Property of Equality (p. 56)

Division Property of Equality (p. 56)

Additional Examples

Example 1

Solve the equation $\frac{h}{2} = 13$. Check your answer.

$$\frac{h}{2} = 13$$

$$(\quad)\frac{h}{2} = 13(\quad) \quad \text{Think: } h \text{ is \_\_\_\_\_ by 2, so \_\_\_\_\_ both sides by 2 to isolate } h.$$

$$\underline{} = \underline{}$$

Check

$$\frac{h}{2} = 13$$

$$\frac{\underline{}}{2} \stackrel{?}{=} 13 \qquad \text{Substitute \_\_\_\_\_ for } h.$$

$$\stackrel{?}{=} 13 \checkmark \qquad \underline{} \text{ is a solution.}$$

LESSON 1-12 CONTINUED

Example 2

Solve the equation 51 = 17x. Check your answer.

$51 = 17x$ Think: x is multiplied by ____ so divide both sides

☐ = ☐   by ____ to ____ x.

☐ = x

Check

$51 = 17x$

$51 \stackrel{?}{=} 17()$ Substitute 3 for ____.

$51 \stackrel{?}{=} 51$ ✓   3 is a ____.

Example 3

Trevor's heart rate is 78 beats per minute. How many times does his heart beat in 10 seconds?

Use the given information to write an _____, where b is the number of heart beats in 10 seconds.

If you count your heart beats for ____ seconds and then multiply that by ____, you can find your heart rate in beats per minute.

Beats in 10s   times 6   = beats per minutes

☐ · ☐ = ☐

b = ☐   Think: b is multiplied by ____, so divide both sides by ____ to isolate the _____.

$\dfrac{6b}{\square} = \dfrac{78}{\square}$

b = ☐

Trevor's heart beats ____ times in 10 seconds.

LESSON 1-12 CONTINUED

Try This

1. Solve the equation $\frac{x}{5} = 30$. Check your answer.

2. Solve the equation $76 = 19y$. Check your answer.

3. During a stock car race, one driver is able to complete 68 laps in 1 hour. How many laps would he finish in 15 minutes?

Chapter Review

1-1 Numbers and Patterns

Identify a possible pattern. Use the pattern to identify the next three numbers or figures.

1. 2, 7, 12, 17, 22, . . .
2. 128, 64, 32, 16, . . .

3. Make a table that shows the number of shaded triangles in each figure. Then tell how many shaded triangles are in the fourth figure. Use drawings to justify your answer.

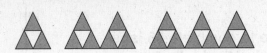

1-2 Exponents

Find each value.

4. 12^2
5. 9^0
6. 5^4
7. 200^1

8. 64, base 2
9. 10,000, base 10
10. 1, base 6
11. 12, base 12

12. Marvin's earnings double each day. He earned $1 on Day 1. How much did Marvin earn on Day 8?

1-3 Metric Measurements

Choose the most appropriate metric unit for each measurement. Justify your answer.

13. The width of your driveway

14. The capacity of a mop bucket

CHAPTER 1 REVIEW CONTINUED

1-4 Applying Exponents

Multiply.

15. 6×10^2 **16.** 13×10^3 **17.** 50×10^1 **18.** $28{,}900 \times 10^4$

Write each number in scientific notation.

19. $64{,}000$ **20.** 300×10^2 **21.** $789{,}000$ **22.** $5{,}000{,}000$

1-5 Order of Operations

Simplify each expression.

23. $2 + 7 \cdot 9 - 4$ **24.** $[(2 + 4)^2 \div 12]^3$ **25.** $4^2 \div (1 + 7)^2 \cdot 12$

Compare. Write <, >, or =.

26. $14 - 2 \cdot 2 \;\blacksquare\; 7 + 6 \div 3$ **27.** $18 \div 2 + 3 \;\blacksquare\; 5 \cdot 4 - 8$

28. $7 + (8 - 4)^2 \;\blacksquare\; 6^2 - 11$ **29.** $19 + 7^2 \div 2 \;\blacksquare\; (4 + 4)^2$

1-6 Properties

Tell which property is represented.

30. $2 \cdot (4 \cdot 3) = (2 \cdot 4) \cdot 3$ **31.** $7 + 5 = 5 + 7$ **32.** $5 \cdot 1 = 5$

Use the Distributive Property to find each product.

33. $5(11 + 7) =$ **34.** $(19 + 3)2 =$

CHAPTER 1 REVIEW *CONTINUED*

1-7 Variables and Algebraic Expressions

Evaluate $n + 8$ for each value of n.

35. $n = 4$ **36.** $n = 0$ **37.** $n = 10$ **38.** $n = 17$

Evaluate each expression for the given value of the variable.

39. $10x + 9$ for $x = 3$ **40.** $14y - 6$ for $y = 5$

41. $2c^2 + 4c$ for $c = 4$ **42.** $z \div 8 + z$ for $z = 40$

Evaluate each expression for the given values of the variables.

43. $\frac{15}{y} + 6z$ for $y = 3$ and $z = 7$ **44.** $12p - 4q + 9$ for $p = 2$ and $q = 1$

1-8 Translate Words Into Math

Write each phrase as an algebraic expression.

45. 12 less than a number **46.** the sum of 6 and a number

47. 5 times a number **48.** 4 divided into a number

49. The Cookie Factory sold q chocolate chip cookies for $0.45 each. Write an algebraic expression for the amount sold.

1-9 Simplifying Algebraic Expressions

Identify like terms in each list.

50. x y^2 x^4 $2y^2$ $\frac{x}{5}$ $4y$ **51.** 4 p^3 $7q$ $2q^2$ $9p^3$ 13

Simplify each expression.

52. $2e + 4f + 7e$ **53.** $b^2 + 2c + 9b^2 + 2$

CHAPTER 1 REVIEW CONTINUED

1-10 Equations and Their Solutions

Determine whether the given value of the variable is a solution.

54. $44 = x - 23$; $x = 21$ **55.** $7n = 42$; $n = 6$ **56.** $y \div 8 = 5$; $y = 40$

57. Which problem situation best matches the equation $15p + 1 = 31$?

Situation A: Bruno's Pizza restaurant charges a $1 delivery fee, plus $15 per pizza. Jeanette paid $31 for pizza. How many pizzas did Jeanette order?

Situation B: Bruno's Pizza Restaurant charges a $1 deliver fee, plus $31 per pizza. Jeanette paid $15 for pizza. How many pizzas did Jeanette order?

1-11 Addition and Subtraction Equations

Solve each equation. Check your answer.

58. $x + 11 = 19$ **59.** $n - 45 = 19$ **60.** $k + 90 = 127$

61. $58 = t + 18$ **62.** $43 = j - 15$ **63.** $w - 13 = 57$

1-12 Multiplication and Division Equations

Solve each equation. Check your answer.

64. $8 = t \div 7$ **65.** $\frac{m}{5} = 25$ **66.** $14 = \frac{p}{7}$

67. $12b = 60$ **68.** $121 = 11a$ **69.** $15x = 135$

CHAPTER 1 Big Ideas

Answer these questions to summarize the important concepts from Chapter 1 in your own words.

1. Explain why 100^5 is greater than 100^4.

2. Explain how to write 45,100,000 in scientific notation.

3. List the steps for the order of operations.

4. Explain how to solve the equation $x - 18 = 35$.

5. Explain how to solve the equation $\frac{x}{9} = 14$.

For more review of Chapter 1:

- Complete the Chapter 1 Study Guide and Review on pages 64–66 of your textbook.
- Complete the Ready to Go On quizzes on pages 32 and 60 of your textbook.

LESSON 2-1 Integers

Lesson Objectives
Compare and order integers and determine absolute value

Vocabulary
opposite (p. 76)

integers (p. 76)

absolute value (p. 77)

Additional Examples

Example 1

Graph the integer −7 and its opposite on a number line.

The opposite of −7 is ____.

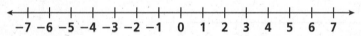

Example 2

Compare the integers. Use < or >.

A. 4 ▨ −4

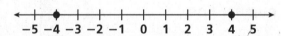

4 is farther to the _____ than −4, so 4 ___ −4.

B. −15 ▨ −9

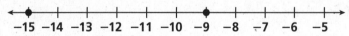

−15 is farther to the _____ than −9, so −15 ___ −9.

LESSON 2-1 CONTINUED

Example 3

Use a number line to order the integers from least to greatest.

−3, 6, −5, 2, 0, −8

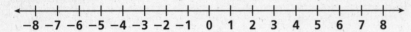

Example 4

Use a number line to find each absolute value.

A. |8|

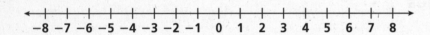

8 is 8 units from 0, so |8| = _____.

B. |−12|

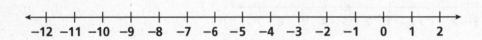

−12 is 12 units from 0, so |−12| = _____.

Try This

1. Graph the integer 1 and its opposite on a number line.

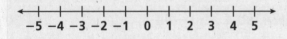

2. Compare the integers. Use < or >.

   −7 ▓ −10

LESSON 2-2: Adding Integers

Lesson Objectives
Add integers

Additional Examples

Example 1

Use a number line to find each sum.

A. $-7 + (-4)$

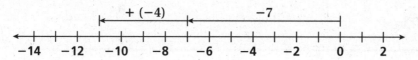

Start at 0. Move left ____ units. Then move left ____ more units.

$-7 + (-4) = $ ____

B. $-12 + 19$

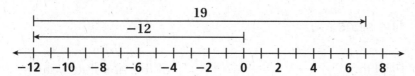

Start at ____. Move ____ 12 units. Then move 19 units.

$-12 + 19 = $ ____

Example 2

Find each sum.

A. $-4 + 8$ The signs are _____.

Find the difference of the _____ values.
Think: $8 - 4 = 4$.

Use the sign of the integer with the _____ absolute value (positive).

37

LESSON 2-2 CONTINUED

Find each sum.

B. $23 + (-35)$ The _____ are different.

$23 + (-35)$ Find the _____ of the absolute values.

Think: ☐ − ☐ = 12.

Use the _____ of the integer with the greater absolute value (_____).

Example 3

Evaluate $x + y$ for $x = -42$, $y = 71$.

$x + y$ _____ for x and y.

☐ + ☐   The signs are _____.

Find the _____ of the absolute values.

Think: ☐ − ☐ = 29.

Use the sign of the integer with the greater absolute _____ (positive).

LESSON 2-2 CONTINUED

Example 4

The jazz band's income from a bake sale was $286. Expenses were $21. Use integer addition to find the band's total profit or loss.

$286 + (-21)$ Use negative for the _____.

$286 - 21$ Find the _____ of the absolute values.

_____   The _____ is positive.

The band's profit was $ _____.

Try This

1. Use a number line to find the sum.

   $-4 + (-5)$

2. Find the sum.

   $-13 + (-24)$

3. Evaluate $x + y$ for $x = -24$, $y = 17$.

4. The French Club was raising money for a trip to Washington D. C. Their carwash raised $730. They had expenses of $52. Use integer addition to find the club's total profit or loss.

LESSON 2-3: Subtracting Integers

Lesson Objectives
Subtract integers

Additional Examples

Example 1

Use a number line to find each difference.

A. $4 - 1$

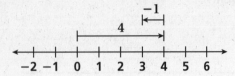

Start at 0. Move right ☐ units. To subtract ☐, move to the left.

$4 - 1 = $ ☐

B. $-3 - 1$

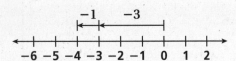

☐ at 0. Move 3 units ☐. To subtract 1, move to the ☐. $-3 - 1 = $ ☐

Example 2

Find each difference.

A. $5 - (-2)$

$5 + 2$ Add the ☐ of -2.

☐

B. $-3 - 7$

☐ + ☐   Add the opposite of 7.

☐

LESSON 2-3 CONTINUED

Example 3

Evaluate $x - y$ for each set of values.

A. $x = -3$ and $y = 2$

$x - y$

$-3 - 2 = $ ____ $+ ($ ____ $)$ for x and y.

$ = $ ____   Add the opposite of ____.

B. $x = 4$ and $y = -6$

$x - y$

$4 - (-6) = $ ____ $+$ ____   Substitute for ____ and ____.

$ = $ ____   Add the ____ of -6.

Example 4

Find the difference between 32°F and −10°F.

____ $- ($ ____ $)$

32 ____ $10 = $ ____   Add the opposite of ____.

The difference in temperature is ____ °F.

Try This

1. Use a number line to find the difference.

$-4 - (-2)$

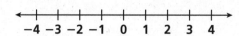

LESSON 2-4: Multiplying and Dividing Integers

Lesson Objectives
Multiply and divide integers

Additional Examples

Example 1

Use a number line to find each product.

A. $-7 \cdot 2$

$-7 \cdot 2 = 2 \cdot (-7)$ Use the Commutative Property.

= ▢   Think: Add -7 ▢ times.

B. $-8 \cdot 3$

$-8 \cdot 3 = 3 \cdot (-8)$ Use the Commutative Property.

= ▢   Think: Add -8 ▢ times.

Example 2

Find each product.

A. $-6 \cdot (-5)$

$-6 \cdot (-5) =$ ▢   Both signs are ▢ , so the product is ▢ .

B. $-4 \cdot 7$

$-4 \cdot 7 =$ ▢   The signs are ▢ , so the product is ▢ .

LESSON 2-4 CONTINUED

Example 3

Find each quotient.

A. $35 \div (-5)$

$35 \div (-5)$ Think: $35 \div \underline{} = 7$.

The \underline{} are different, so the \underline{} is negative.

B. $-32 \div (-8)$

$-32 \div (-8)$ Think: $32 \div 8 = 4$.

The signs are \underline{}, so the quotient is \underline{}.

Example 4

Mrs. Johnson kept track of a stock she was considering buying. She recorded the price change each day. What was the average change per day?

Mon	Tues	Wed	Thu	Fri
−$1	$3	$2	−$5	$6

Find the \underline{} of the changes in price.

$\dfrac{5}{5} =$ Divide to find the average.

The average change was $\underline{}$ per day.

Try This

1. Use a number line to find the product.

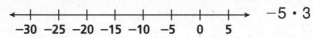 $-5 \cdot 3$

2. Find the product.

$-2 \cdot (-8)$

LESSON 2-5: Solving Equations Containing Integers

Lesson Objectives

Solve one-step equations with integers

Additional Examples

Example 1

Solve each equation.

A. $-6 + x = -7$

    +____   +____   Add ____ to both sides to isolate the variable.

    $x =$ ____

B. $p + 5 = -3$

    +(____) +(____)   Add ____ to both sides to ____ the variable.

    $p =$

C. $y - 9 = -40$

    +____   +____   Add ____ to both sides to ____ the variable.

    $y =$

Example 2

Solve each equation. Check each answer.

A. $\dfrac{b}{-5} = 6$

$\dfrac{b}{-5} = 6$

$(\_\_\_)\dfrac{b}{-5} = (\_\_\_)6$ Multiply both sides by ____ to isolate the

$b =$ ____

LESSON 2-5 CONTINUED

B. $-400 = 8y$

$-400 = 8y$

$\dfrac{-400}{\boxed{}} = \dfrac{8y}{\boxed{}}$ Divide both sides by $\boxed{}$ to $\boxed{}$ the variable.

$\boxed{} = y$

Example 3

In 2003, a manufacturer made a profit of $300 million. This amount was $100 million more than the profit in 2002. What was the profit in 2002?

Let p represent the profit in 2002 (in millions of dollars).

Profit in 2003 = $p + \boxed{}$

Profit in 2003 = $\$\boxed{}$ million

$p + 100 = 300$

$\underline{-\boxed{}} \underline{-\boxed{}}$

$p = \boxed{}$

The profit in 2002 was $\$\boxed{}$ million.

Try This

1. Solve the equation. Check the answer.

   $y - 7 = -34$

2. Solve the equation. Check the answer.

   $\dfrac{c}{4} = -24$

LESSON 2-6 Prime Factorization

Lesson Objectives
Find the prime factorizations of composite numbers

Vocabulary
prime number (p. 106) _____

composite number (p. 106) _____

prime factorization (p. 106) _____

Additional Examples

Example 1

Tell whether each number is prime or composite.

A. 11

The factors of 11 are ____ and ____.

So 11 is _____.

B. 16

The factors of 16 are ____, ____, ____, ____, and ____.

So 16 is _____.

LESSON 2-6 CONTINUED

Example 2

Write the prime factorization of each number.

A. 24

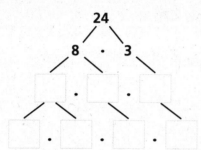

Write _____ as the product of two factors.

Continue factoring until all factors are _____.

The prime factorization of 24 is _____. Using exponents, you can write this as _____.

B. 150

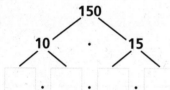

Write _____ as the product of two factors. Continue factoring until all factors are _____.

The prime factorization of 150 is _____, or _____.

LESSON 2-6 CONTINUED

Example 3

Write the prime factorization of each number.

A. 476

```
2 | 476
2 | 238
7 | 119
17 | 17
     1
```

Divide 476 by 2. Write the quotient below 476.

Keep dividing by a _____ number.

Stop when the quotient is ☐.

The prime factorization of 476 is _____, or _____.

B. 275

```
5 | 275
5 | 55
11 | 11
     1
```

Divide 275 by 5. Write the quotient below 275.

Keep dividing by a _____ factor.

Stop when the quotient is ☐.

The prime factorization of 275 is _____, or _____.

Try This

1. Tell whether the number is prime or composite.

39

2. Write the prime factorization of the number.

90

3. Write the prime factorization of the number.

325

LESSON 2-7 Greatest Common Factor

Lesson Objectives

Find the greatest common factor of two or more whole numbers

Vocabulary

greatest common factor (GCF) (p. 110)

Additional Examples

Example 1

Find the greatest common factor (GCF) of 12, 36, and 54.

12: 1, 2, 3, 4, 6, 12
36: 1, 2, 3, 4, 6, 9, 12, 18, 36
54: 1, 2, 3, 6, 9, 18, 27, 54

List all of the _____ of each number.

Circle the _____ factor that is in all the lists.

The GCF is _____.

Example 2

Find the greatest common factor (GCF).

A. 40, 56

$40 = 2 \cdot 2 \cdot 2 \cdot 5$
$56 = 2 \cdot 2 \cdot 2 \cdot 7$

$2 \cdot 2 \cdot 2 =$ _____

The GCF is _____.

Write the _____ factorization of each number and circle the _____ prime factors.

_____ the common prime factors.

LESSON 2-7 CONTINUED

Find the greatest common factor (GCF).

B. 252, 180, 96, 60

$252 = 2 \cdot 2 \cdot 3 \cdot 3 \cdot 7$
$180 = 2 \cdot 2 \cdot 3 \cdot 3 \cdot 5$
$96 = 2 \cdot 2 \cdot 2 \cdot 2 \cdot 2 \cdot 3$
$60 = 2 \cdot 2 \cdot 3 \cdot 5$

Write the prime _____ of each number and circle the common _____ factors.

☐ · ☐ · ☐ = 12   Multiply the common prime _____.

The GCF is _____.

Example 3

PROBLEM SOLVING APPLICATION

You have 120 red beads, 100 white beads, and 45 blue beads. You want to use all the beads to make bracelets that have red, white, and blue beads on each. What is the greatest number of bracelets you can make?

1. **Understand the Problem**
   Rewrite the question as a statement.

   • Find the _____ number of bracelets you can make.

   List the important information:

   • There are _____ red beads, _____ white beads, and _____ blue beads.

   • Each bracelet must have the _____ number of red, white, and blue beads.

   The answer will be the _____ of 120, 100, and 45.

2. **Make a Plan**
   You can list the prime factors of 120, 100, and 45 to find the GCF.

LESSON 2-7 CONTINUED

3. Solve

120 = ☐ · ☐ · ☐ · ☐ · ☐

100 = ☐ · ☐ · ☐ · ☐

45 = ☐ · ☐ · ☐

The GCF of 120, 100, and 45 is ☐.

You can make ☐ bracelets.

4. Look Back
If you make 5 bracelets, each one will have ☐ red beads, ☐ white beads, and ☐ blue beads, with nothing left over.

Try This

1. Find the greatest common factor (GCF) of 14, 28, and 63.

2. Find the greatest common factor (GCF).
   360, 250, 170, 40

3. Nathan has made fishing flies that he plans to give away as gift sets. He has 24 wet flies and 18 dry flies. Using all of the flies, how many sets can he make?

Holt Mathematics

LESSON 2-8 Least Common Multiple

Lesson Objectives

Find the least common multiple of two or more whole numbers

Vocabulary

multiple (p. 114) _____

_____

least common multiple (LCM) (p. 114) _____

_____

Additional Examples

Example 1

Find the least common multiple (LCM).

A. 2, 7

2: 2, 4, 6, 8, 10, 12, 14   List some _____ of each number.

7: 7, 14, 21, 28, 35   Find the _____ value that is in both lists.

The LCM is ____.

B. 3, 6, 9

3: 3, 6, 9, 12, 15, 18, 21   List some _____ of each number.

6: 6, 12, 18, 24, 30   Find the _____ value that is in all the lists.

9: 9, 18, 27, 36, 45

The LCM is ____.

LESSON 2-8 CONTINUED

Example 2

Find the least common multiple (LCM).

A. 60, 130

60 = 2 · 2 · 3 · 5   Write the _____ factorization of each number.

130 = 2 · 5 · 13   Circle the _____ prime factors.

☐ , ☐ , ☐ , ☐ , ☐   List the _____ factors, using the circled factors only once.

☐ · ☐ · ☐ · ☐   _____ the factors in the list.

The LCM is ☐.

B. 14, 35, 49

14 = 2 · 7   Write the prime _____ of each number.
35 = 5 · 7
49 = 7 · 7   Circle the common prime _____.

☐ , ☐ , ☐   List the prime _____, using the circled factors only once.

☐ · ☐ · ☐   Multiply the _____ in the list.

The LCM is ☐.

LESSON 2-8 CONTINUED

Example 3

Mr. Washington will set up the band chairs all in rows of 6 or all in rows of 8. What is the least number of chairs he will set up?

Find the LCM of ____ and ____.

6 = ☐ · ☐

8 = ☐ · ☐ · ☐

The LCM is ☐ · ☐ · ☐ · ☐ = ☐.

He will set up at least ____ chairs.

Try This

1. Find the least common multiple (LCM).

    2, 6, 4

2. Find the least common multiple (LCM).

    18, 36, 54

3. Two satellites are put into orbit over the same location at the same time. One orbits the earth every 24 hours, while the second completes an orbit every 18 hours. How much time will elapse before they are once again over the same location at the same time?

LESSON 2-9: Equivalent Fractions and Mixed Numbers

Lesson Objectives
Identify, write, and convert between equivalent fractions and mixed numbers

Vocabulary

equivalent fractions (p. 120) _____

improper fractions (p. 121) _____

mixed number (p. 121) _____

Additional Examples

Example 1

Find two fractions equivalent to $\frac{5}{7}$.

$\frac{5}{7} = \frac{5 \cdot \boxed{}}{7 \cdot \boxed{}} = \boxed{}$ Multiply numerator and denominator by $\boxed{}$.

$\frac{5}{7} = \frac{5 \cdot \boxed{}}{7 \cdot \boxed{}} = \boxed{}$ Multiply numerator and denominator by $\boxed{}$.

Example 2

Write the fraction $\frac{18}{24}$ in simplest form.

Find the GCF of $\boxed{}$ and $\boxed{}$.

$18 = 2 \cdot 3 \cdot 3$
$24 = 2 \cdot 2 \cdot 2 \cdot 3$

The GCF is $\boxed{} = 2 \cdot 3$.

$\frac{18}{24} = \frac{18 \div \boxed{}}{24 \div \boxed{}} = \boxed{}$ Divide the numerator and denominator by $\boxed{}$.

Holt Mathematics

LESSON 2-9 CONTINUED

Example 3

Determine whether the fractions in each pair are equivalent.

A. $\dfrac{4}{6}$ and $\dfrac{28}{42}$

Both fractions can be written with a _____ of 3.

$\dfrac{4}{6} = \dfrac{4 \div \square}{6 \div \square} = \square$

$\dfrac{28}{42} = \dfrac{28 \div \square}{42 \div \square} = \square$

The numerators are equal, so the fractions are _____.

B. $\dfrac{6}{10}$ and $\dfrac{20}{25}$

Both fractions can be written with a demoninator of _____.

$\dfrac{6}{10} = \dfrac{6 \cdot \square}{10 \cdot \square} = \square$

$\dfrac{20}{25} = \dfrac{20 \cdot \square}{25 \cdot \square} = \square$

The numerators are not equal, so the fractions are _____.

LESSON 2-9 CONTINUED

Example 4

A. Write $\frac{13}{5}$ as a mixed number.

First divide the _____ by the _____.

$\frac{13}{5}$ = _____   Use the _____ and _____ to write a mixed number.

B. Write $7\frac{2}{3}$ as an improper fraction.

First multiply the _____ and _____ number, and then add the _____.

$7\frac{2}{3} = \frac{3 \cdot \_\_\_ + \_\_\_}{3}$ = _____   Use the result to write the improper fraction.

Try This

1. Find two fractions equivalent to $\frac{3}{8}$.

2. Write the fraction $\frac{24}{36}$ in simplest form.

3. Determine whether the fractions in each pair are equivalent.

   $\frac{3}{9}$ and $\frac{6}{18}$

4. Write $\frac{15}{6}$ as a mixed number.

LESSON 2-10: Equivalent Fractions and Decimals

Lesson Objectives

Write fractions as decimals, and vice versa, and determine whether a decimal is terminating or repeating

Vocabulary

terminating decimal (p. 124) _____

_____

repeating decimal (p. 124) _____

Additional Examples

Example 1

Write each fraction as a decimal. Round to the nearest hundredth, if necessary.

A. $\frac{1}{4}$

B. $\frac{9}{5}$

C. $\frac{5}{3}$

```
    _____              _____              _____
  4)1.00              5)9.0               3)5.00
   -8                  -5                  -3
   ---                 ---                 ---
    20                  40                  20
   -20                 -40                 -18
   ---                 ---                 ---
     0                   0                  20
                                           -18
                                           ---
                                             2
```

$\frac{1}{4} = $ _____ $\frac{9}{5} = $ _____ $\frac{5}{3} \approx $ _____

Example 2

Write each fraction as a decimal.

A. $\frac{4}{5}$

$\frac{4}{5} \times \frac{\square}{\square} = \frac{\square}{\square}$ Multiply to get a power of _____ in the denominator.

= _____

LESSON 2-10 CONTINUED

B. $\frac{37}{50}$

$\frac{37}{50} \times \underline{} = \underline{}$ Multiply to get a power of ___ in the denominator.

$= \underline{}$

Example 3

Write each decimal as a fraction in simplest form.

A. 0.018

$0.018 = \frac{18}{1{,}000} = \frac{18 \div 2}{1{,}000 \div 2} = \underline{}$

B. 1.55

$1.55 = \frac{155}{100} = \frac{155 \div 5}{100 \div 5} = \underline{}$ or $\underline{}$

Example 4

A football player completed 1,546 of the 3,875 passes he attempted. Find his completion rate. Write your answer as a decimal rounded to the nearest thousandth.

Fraction	What the Calculator Shows	Completion Rate
	1546 ÷ 3875 ENTER 0.398967749	

His completion rate is _____.

Try This

1. Write the fraction as a decimal. Round to the nearest hundredth, if necessary.

   $\frac{6}{5}$

LESSON 2-11: Comparing and Ordering Rational Numbers

Lesson Objectives

Compare and order fractions and decimals

Vocabulary

rational number (p. 129)

Additional Examples

Example 1

Compare the fractions. Write < or >.

A. $\dfrac{7}{9}$ ☐ $\dfrac{5}{8}$

Both fractions can be written with a ☐ of 72.

$\dfrac{7}{9} = \dfrac{7 \cdot \Box}{9 \cdot \Box} = \Box$ Write as fractions with ☐ denominators.

$\dfrac{5}{8} = \dfrac{5 \cdot \Box}{8 \cdot \Box} = \Box$

☐ > ☐, and so $\dfrac{7}{9}$ ☐ $\dfrac{5}{8}$. Compare the numerators.

B. $-\dfrac{2}{5}$ ☐ $-\dfrac{3}{7}$

Both fractions can be written with a ☐ of 35.

$-\dfrac{2}{5} = \dfrac{-2 \cdot \Box}{5 \cdot \Box} = \Box$ Write as fractions with common ☐.

$-\dfrac{3}{7} = \dfrac{-3 \cdot \Box}{7 \cdot \Box} = \Box$ Put the negative signs in the numerators.

$-\Box$ > $-\Box$, and so $-\dfrac{2}{5}$ ☐ $-\dfrac{3}{7}$.

LESSON 2-11 CONTINUED

Example 2

Compare the decimals. Write < or >.

A. 0.427 ▓ 0.425

0.427   Line up the _____ points.

The tenths and hundredths are the _____.

0.425   Compare the _____ : 7 ☐ 5.

0.427 ☐ 0.425

B. $0.7\overline{3}$ ▓ 0.734   _____ is a repeating decimal.

$0.7\overline{3}$ = 0.733   Line up the _____ points.

The tenths and hundredths are the _____.

0.734   Compare the _____ : 3 ☐ 4.

$0.7\overline{3}$ ☐ 0.734

LESSON 2-11 CONTINUED

Example 3

Order the numbers from least to greatest.

$\frac{4}{5}$, 0.93, and 0.9

Write as decimals with the same number of places.

$\frac{4}{5} = 0.80$ 0.93 = ____   0.9 = ____

Graph the numbers on a number line.

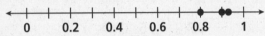

The values on a number line _____ as we move from left to right.

0.80 < 0.90 < 0.93   Place the decimals in _____.

Try This

1. Compare the fractions. Write < or >.

   $\frac{5}{6}$ ▪ $\frac{7}{8}$

   $\frac{5}{6}$ ____ $\frac{7}{8}$

2. Compare the decimals. Write < or >.

   0.535 ▪ 0.538

   0.535 ____ 0.538

3. Order the numbers from least to greatest.

   $\frac{3}{5}$, 0.84, and 0.7

Chapter 2 Chapter Review

2-1 Integers

Compare the integers. Use < or >.

1. 4 ▪ −6
2. −2 ▪ 2
3. −14 ▪ −9

Use a number line to find each absolute value.

4. $|-6|$
5. $|3|$
6. $|-9|$

2-2 Adding Integers

Find each sum.

7. $-8 + 6$
8. $13 + (-3)$
9. $-7 + (-4)$

Evaluate $a + b$ for the given values.

10. $a = 8, b = -17$
11. $a = -44, b = 49$
12. $a = -5, b = -14$

2-3 Subtracting Integers

Find each difference.

13. $7 - 11$
14. $-9 - (-15)$
15. $-7 - 6$

Evaluate $a - b$ for the given values.

16. $a = 8, b = -3$
17. $a = -4, b = 11$
18. $a = 3, b = 8$

2-4 Multiplying and Dividing Integers

Find each product.

19. $6 \cdot (-3)$
20. $-4 \cdot 8$
21. $-8 \cdot (-5)$

Find each quotient.

22. $45 \div (-9)$
23. $-24 \div 4$
24. $-39 \div (-3)$

CHAPTER 2 REVIEW CONTINUED

2-5 Solving Equations Containing Integers

Solve. Check your answer.

25. $8h = -48$ **26.** $-15 + k = 35$ **27.** $27 - x = -42$

28. This year, 84 students performed at the Spring Choral Concert. There were 4 groups with an equal number of students that performed. How many students were in each group?

2-6 Prime Factorization

Tell whether each number is prime or composite.

29. 23   **30.** 35   **31.** 46

Write the prime factorization of each number.

32. 75   **33.** 48   **34.** 63

2-7 Greatest Common Factor

Find the greatest common factor (GCF).

35. 35, 49   **36.** 36, 48, 60   **37.** 54, 18, 72, 36

38. A Sports Club is preparing welcome gifts. There are 65 golf balls, 39 baseballs, and 26 tennis balls. What is the greatest number of gifts the Sports Club can prepare using all of the golf balls, baseballs, and tennis balls?

2-8 Least Common Multiple

Find the least common multiple (LCM).

39. 8, 10   **40.** 3, 8, 9   **41.** 2, 3, 4, 6

42. Charlie and Heather swam laps in the pool. Charlie completed one lap in 3 minutes and Heather completed one lap in 7 minutes. If they started swimming laps at the same time, in how many minutes will they finish a lap together?

CHAPTER 2 REVIEW *CONTINUED*

2-9 Equivalent Fractions and Mixed Numbers

Write each fraction as an improper fraction or a mixed number.

43. $2\frac{13}{14}$ 44. $\frac{17}{6}$ 45. $\frac{15}{11}$ 46. $3\frac{2}{7}$

Determine whether the fractions in each pair are equivalent.

47. $\frac{3}{7}$ and $\frac{4}{9}$ 48. $\frac{9}{6}$ and $\frac{12}{8}$ 49. $\frac{4}{5}$ and $\frac{12}{15}$

2-10 Equivalent Fractions and Decimals

Write each fraction as a decimal.

50. $\frac{2}{5}$ 51. $\frac{7}{20}$ 52. $\frac{37}{50}$ 53. $\frac{5}{8}$

Write each decimal as a fraction in simplest form.

54. 0.075 55. −1.15 56. 0.38 57. −2.8

2-11 Comparing and Ordering Rational Numbers

Compare the fractions. Write < or >.

58. $\frac{3}{5}$ ■ $\frac{2}{3}$ 59. $-\frac{6}{5}$ ■ $-\frac{5}{6}$ 60. $-\frac{4}{7}$ ■ $-\frac{5}{9}$

Compare the decimals. Write < or >.

61. 0.378 ■ 0.375 62. −0.19 ■ −0.919 63. −5.9 ■ 5.09

CHAPTER 2 Big Ideas

Answer these question to summarize the important concepts from Chapter 2 in your own words.

1. Explain why −5 and 5 have the same absolute value.

2. Explain how to find the sign of the answer when subtracting integers.

3. Explain how to find the sign of the answer when multiplying or dividing integers.

4. Explain how to find the GCF of two or more whole numbers.

5. Explain why $0.\overline{7}$ is a rational number.

For more review of Chapter 2:

- Complete the Chapter 2 Study Guide and Review on pages 138–140 of your textbook.
- Complete the Ready to Go On quizzes on pages 104, 118, and 132 of your textbook.

LESSON 3-1: Estimate with Decimals

Lesson Objectives

Estimate decimal sums, differences, products, and quotients

Vocabulary

compatible numbers (p. 150)

Additional Examples

Example 1

Estimate by rounding to the nearest integer.

A. $4.5 + 8.9$

    $4.5 \longrightarrow \boxed{}$ $5 \geq 5$, so round to ____ .

    $+8.9 \longrightarrow +\underline{}$ $9 > 5$, so round to ____ .

    $\boxed{} \longleftarrow$ Estimate

B. $28.3 - 11.7$

    $28.3 \longrightarrow \boxed{}$ $3 < 5$, so round to ____ .

    $-11.7 \longrightarrow -\underline{}$ $7 > 5$, so round to ____ .

    $\boxed{} \longleftarrow$ Estimate

Example 2

Use compatible numbers to estimate.

A. $45.99 \cdot 2.31$

    $45.99 \longrightarrow \boxed{}$ Round to the nearest multiple of 10.

    $\times 2.31 \longrightarrow \times \underline{}$ $3 < 5$, so round to ____ .

    $\boxed{} \longleftarrow$ Estimate

LESSON 3-1 CONTINUED

B. 51.33 ÷ (−7.98)

51.33 ⟶ ☐   51 is prime, so round to ☐.

−7.98 ⟶ ☐   −7 divides into 49 without a ☐.

49 ÷ −7 = ☐   ⟵ **Estimate**

Example 3

Cara spent $58.80 on 4.8 pounds of lobster. Use estimation to check whether $12 per pound is reasonable to spend on lobster.

$58.80 ⟶ $ ☐   to the nearest multiple of 10.

4.85 ⟶ ☐   8 > 5, so round to ☐.

60 ÷ 5 = ☐   ⟵ **Estimate.**

The estimate is ☐.

Try This

1. Estimate by rounding to the nearest integer.

19.2 − 13.6

2. Use compatible numbers to estimate.

19.42 ÷ (−4.88)

LESSON 3-2: Adding and Subtracting Decimals

Lesson Objectives

Add and subtract decimals

Additional Examples

Example 1

Add. Estimate to check whether each answer is reasonable.

A. $4.55 + 11.3$

    4.55   Line up the _____ points.

  + 11.30   Use _____ as a placeholder.

  _____   Add.

Estimate

  $5 + 11 =$ _____   _____ is a reasonable answer.

B. $-8.33 + (-10.972)$

  $-8.33 + (-10.972)$ Think: _____ + _____

    8.330   Line up the _____ points.

  + 10.972   Use zero as a _____.

  _____   Add.

  $-8.33 + (-10.972) =$ _____

Estimate

  _____ + _____ = -19 -19.302 is a _____ answer.

LESSON 3-2 CONTINUED

Example 2

Subtract.

A. 5.34 − 2.08

```
  5.34
− 2.08
```
up the decimal points.

Subtract.

B. 28 − 15.911

```
   7 9 910
  28.000      Use zeros as placeholders.
− 15.911      Line up the decimal points.
```

Subtract.

Example 3

During one month in the United States, 492.23 million commuter trips were taken on buses, and 26.331 million commuter trips were taken on light rail. What was the total number of trips taken on buses and light rail? Estimate to check whether your answer is reasonable.

```
   492.230          Line up the decimal
 +  26.331          Use          as a placeholder.
                    Add.
```

Estimate

☐ + ☐ = 518   ☐ is a reasonable answer.

The total number of trips was ☐ million.

Try This

1. Add. Estimate to check whether the answer is reasonable.

   −7.89 + (−13.852)

LESSON 3-3: Multiplying Decimals

Lesson Objectives
Multiply decimals

Additional Examples

Example 1

Multiply.

A. 7 · 0.1

7	0 decimal places
× 0.1	1 decimal place
☐	0 + 1 = 1 decimal place.

B. −3 · 0.03

−3	0 decimal places
× 0.03	2 decimal places
☐	0 + 2 = 2 decimal places. Use ☐ as a placeholder.

C. 2.45 · 35

2.45	☐ decimal places
× 35	☐ decimal places
☐	☐ + ☐ = ☐ decimal places.

Example 2

Multiply. Estimate to check whether each answer is reasonable.

A. 2.4 · 1.8

```
    2.4      1 decimal place
  × 1.8      1 decimal place
  ─────
   1 92
 + 2 40
  ─────
   ☐         1 + 1 = 2 decimal places.
```

Estimate

2 · 2 = 4   ☐ is a reasonable answer.

LESSON 3-3 CONTINUED

Multiply. Estimate to check whether each answer is reasonable.

B. $-3.84 \cdot 0.9$

$$\begin{array}{r} -3.84 \\ \times\ 0.9 \\ \hline \end{array}$$

2 decimal places
1 decimal place

$2 + 1 = 3$ decimal places.

Estimate

$-4 \cdot 1 = -4$ ☐ is a ☐ answer.

Example 3

To find your weight on another planet, multiply the relative gravitational pull of the planet and your weight. The relative gravitational pull on Mars is 0.38. What would a person who weighs 85 pounds on Earth weigh on Mars?

$$\begin{array}{r} 85 \\ \times\ 0.38 \\ \hline 680 \\ +\ 2550 \\ \hline \end{array}$$

☐ decimal places
☐ decimal places
☐ + ☐ = ☐ decimal places

Estimate

$85 \times 0.5 =$ ☐   ☐ is a reasonable answer.

The person would weigh _____ pounds on Mars.

Try This

1. Multiply.

$3.65 \cdot 15$

LESSON 3-4: Dividing Decimals by Integers

Lesson Objectives
Divide decimals by integers

Additional Examples

Example 1

Divide. Estimate to check whether each answer is reasonable.

A. 36.75 ÷ 7

```
  7) 36.75
    -35 ↓
      1 7
     -1 4 ↓
        3 5
       -3 5
          0
```

Place the _____ point in the quotient directly above the decimal point in the dividend.

Divide as with whole numbers.

Estimate

35 ÷ 7 = 5   _____ is a reasonable answer.

B. 0.87 ÷ 3

```
  3) 0.87
    -6
     27
    -27
      0
```

Place the decimal _____ in the quotient directly above the decimal point in the dividend.

Divide as with whole numbers.

Estimate

0.9 ÷ 3 = 0.3   0.29 is a _____ answer.

LESSON 3-4 CONTINUED

Example 2

You can buy juice by the bottle or case. Either way, it costs the same for each bottle. A case of 24 bottles of juice costs $23.52. Kevin bought a bag of peanuts for 75¢ and one bottle of juice. How much did Kevin spend in all?

First find the cost for one bottle of juice by dividing the cost of a case by the number of bottles in a case. Then add the cost of a bag of peanuts.

```
    0.98
24)23.52
   21 6
   ----
    1 92
   -1 92
   ----
       0
```

Place the _____ point in the quotient directly above the decimal point in the dividend.

$0.98 + $0.75 = $ _____   One bottle of juice costs $ _____ and a bag of peanuts costs $ _____.

Kevin spent a total of $ _____.

Try This

1. Divide. Estimate to check whether the answer is reasonable.

   65.16 ÷ (−12)

2. Cookies at a store sell for $1.80 a dozen. The cost for each cookie is the same whether you buy them individually or by the dozen. John decided to buy 1 cookie and a quart of milk. The milk cost $1.79. How much did John have to pay?

LESSON 3-5: Dividing Decimals and Integers by Decimals

Lesson Objectives

Divide decimals and integers by decimals

Additional Examples

Example 1

Divide.

A. $8.28 \div 4.6$

$8.28 \div 4.6 = 82.8 \div 46$ Multiply both numbers by _____ to make the divisor an integer.

```
    ____
46)82.8
   46
   ───
   36 8
  −36 8
   ────
      0
```
Divide as with whole numbers.

B. $18.48 \div (-1.75)$

$18.48 \div (-1.75) = 1{,}848 \div 175$ Multiply both numbers by _____ to make the divisor an integer.

```
     _____
175)1848.00
   −175
   ─────
     98 0
    −87 5
    ─────
    10 50
   −10 50
   ──────
        0
```
Use _____ as placeholders.
Divide as with whole numbers.

$18.48 \div (-1.75) =$ _____   The signs are _____.

LESSON 3-5 CONTINUED

Example 2

Divide. Estimate to check whether each answer is reasonable.

A. $4 \div 1.25$

$4.00 \div 1.25 = 400 \div 125$ 　　　　　　　both numbers by 100.

```
    _____
125)400.0          Use zero as a _____.
   -375            Divide as with whole numbers.
   ____
    25 0
   -25 0
   ____
       0
```

Estimate

$4 \div 1 = 4$ 　　　　　　　The answer is _____.

B. $-24 \div (-2.5)$

$-24.0 \div (-2.5) = -240 \div (-25)$ 　　Multiply both numbers by _____.

```
   _____
25)240.0           Divide as with whole numbers.
   225
   ___
    15 0
   -15 0
   ____
       0
```

Estimate

$-24 \div (-3) = 8$ 　　　　　The answer is _____.

LESSON 3-5 CONTINUED

Example 3

Eric paid $229.25 to rent a car. The fee to rent the car was $32.75 per day. For how long did Eric rent the car?

229.25 ÷ 32.75 = 22,925 ÷ 3,275   Multiply both numbers by ____.

3,275)‾22,925
   −22,925
        0   Divide as with whole numbers.

Eric rented the car for ____ days.

Try This

1. Divide.

   6.45 ÷ 0.5

2. Divide. Estimate to check whether the answer is reasonable.

   −22 ÷ (−2.5)

3. Jace took a trip in which he drove 350 miles. During the trip his truck used 12.5 gallons of gas. What was his truck's gas mileage?

LESSON 3-6: Solving Equations Containing Decimals

Lesson Objectives
Solve one-step equations that contain decimals

Additional Examples

Example 1

Solve.

A. $n - 2.75 = 8.3$

$n - 2.75 = 8.30$

+ _____ + _____   Add to isolate ____.

$n = $ ____

B. $a + 32.66 = 42$

$a + 32.66 = 42.00$

− _____ − _____   Subtract to ____ a.

$a = $ ____

Example 2

Solve.

A. $\dfrac{x}{4.8} = 5.4$

$\dfrac{x}{4.8} = 5.4$

$\dfrac{x}{4.8} \cdot $ ____ $= 5.4 \cdot $ ____   Multiply to ____ x.

$x = $ ____

LESSON 3-6 CONTINUED

Solve.

B. $9 = 3.6d$

$9 = 3.6d$

$\dfrac{9}{\boxed{}} = \dfrac{3.6d}{\boxed{}}$ Divide to isolate $\boxed{}$.

$\boxed{} = d$ Think: $9 \div 3.6 = 90 \div 36$

$\boxed{} = d$

Example 3

PROBLEM SOLVING APPLICATION

A board-game box is 2.5 inches tall. A toy store has shelving space measuring 15 inches vertically in which to store the boxes. How many boxes can be stacked in the space?

1. **Understand the Problem**
   Rewrite the question as a statement.

   Find the number of boxes that can be placed on the shelf.
   List the important information:

   A. Each board-game box is $\boxed{}$ inches tall.

   B. The store has shelving space measuring $\boxed{}$ inches.

2. **Make a Plan**

   The total height of the boxes is equal to the height of one box $\boxed{}$ the number of boxes. Since you know how tall the shelf is you can write an equation with b being the number of boxes.

   $\boxed{} \, b = 15$

LESSON 3-6 CONTINUED

3. Solve

$2.5b = 15$

$\dfrac{2.5b}{\boxed{}} = \dfrac{2.5}{\boxed{}}$ Divide to $\boxed{}$ b.

$b = \boxed{}$

$\boxed{}$ boxes can be stacked in the space.

4. Look Back

You can round 2.5 to $\boxed{}$ and estimate how many boxes will fit on the shelf.

$15 \div \boxed{} = 5$

So $\boxed{}$ boxes is a reasonable answer.

Try This

1. Solve.

$a + 27.51 = 36$

2. Solve.

$9 = 2.5d$

LESSON 3-6 CONTINUED

3. PROBLEM SOLVING APPLICATION

A canned good is 4.5 inches tall. A grocery store has shelving space measuring 18 inches vertically in which to store the cans. How many cans can be stacked in the space?

1. **Understand the Problem**
   Rewrite the question as a statement.

   Find the number of cans that can be placed on the shelf.
   List the important information:

   A. Each can is ____ inches tall.

   B. The store has shelving space measuring ____ inches.

2. **Make a Plan**

   The total height of the cans is equal to the height of one can ____ the number of cans. Since you know how tall the shelf is you can write an equation with c being the number of cans.

   ____ $c = 18$

3. **Solve**

   $4.5c = 18$

   $\dfrac{4.5c}{\_\_} = \dfrac{18}{\_\_}$ Divide to ____ c.

   $c = $ ____

   ____ cans can be stacked in the space.

4. **Look Back**

   You can round 4.5 to ____ and 18 to ____ and estimate how many cans will fit on the shelf.

   ____ ÷ ____ = 4

   So ____ cans is a reasonable answer.

LESSON 3-7: Estimate with Fractions

Lesson Objectives
Estimate sums, differences, products, and quotients of fractions and mixed numbers

Additional Examples

Example 1

A blue whale can grow to $33\frac{3}{5}$ m long, while the great white shark may be as long as $4\frac{1}{2}$ m. Estimate how much longer the blue whale is than the great white shark.

$33\frac{3}{5} - 4\frac{1}{2}$

$33\frac{3}{5} \rightarrow \boxed{}$ $4\frac{1}{2} \rightarrow \boxed{}$ $\boxed{}$ each mixed number.

$\boxed{} - \boxed{} = \boxed{}$ Subtract.

The blue whale is about $\boxed{}$ m longer.

Example 2

Estimate each sum or difference.

A. $\frac{7}{9} - \frac{2}{5}$

$\frac{7}{9} \rightarrow 1$ $\frac{2}{5} \rightarrow \frac{1}{2}$ $\boxed{}$ each fraction.

$\boxed{} - \boxed{} = \boxed{}$ Subtract.

B. $4\frac{5}{9} + 3\frac{1}{8}$

$4\frac{5}{9} \rightarrow 4\frac{1}{2}$ $3\frac{1}{8} \rightarrow 3$ Round each mixed $\boxed{}$.

$\boxed{} + \boxed{} = \boxed{}$ Add.

C. $-2\frac{1}{8} + \frac{7}{12}$

$-2\frac{1}{8} \rightarrow -2$ $\frac{7}{12} \rightarrow \frac{1}{2}$ $\boxed{}$ each number.

$\boxed{} + \boxed{} = \boxed{}$ Add.

Holt Mathematics

LESSON 3-7 CONTINUED

Example 3

Estimate each product or quotient.

A. $3\frac{2}{9} \cdot 6\frac{5}{6}$

$3\frac{2}{9} \rightarrow \square \quad 6\frac{5}{6} \rightarrow \square$ 　　　□ each mixed number to the nearest □.

$\square \cdot \square = \square$ 　　Multiply.

B. $13\frac{4}{5} \div 2\frac{1}{4}$

$13\frac{4}{5} \rightarrow \square \quad 2\frac{1}{4} \rightarrow \square$ 　　　□ each mixed number to the nearest □.

$\square \div \square = \square$ 　　Divide.

Try This

1. A Cocker Spaniel may grow to weigh about $11\frac{1}{2}$ kilograms while the Chihuahua will not weigh more than $2\frac{7}{8}$ kilograms. Estimate how much more a Cocker Spaniel weighs than a Chihuahua.

2. Estimate the sum.

   $-4\frac{2}{9} + \frac{8}{15}$

3. Estimate the quotient.

   $18\frac{2}{3} \div 3\frac{1}{5}$

LESSON 3-8: Adding and Subtracting Fractions

Lesson Objectives
Add and subtract fractions

Additional Examples

Example 1

Add or subtract. Write each answer in simplest form.

A. $\frac{5}{8} + \frac{1}{8}$

$\frac{5}{8} + \frac{1}{8} = \frac{\boxed{} + \boxed{}}{8}$ Add the _____ and keep the _____.

$= \boxed{} = \boxed{}$ Simplify.

B. $\frac{9}{11} - \frac{4}{11}$

$\frac{9}{11} - \frac{4}{11} = \frac{\boxed{} - \boxed{}}{11}$ Subtract the _____ and keep the _____.

$= \boxed{}$ The answer is in the simplest form.

Example 2

Add or subtract. Write each answer in simplest form.

A. $\frac{5}{6} + \frac{7}{8}$

$\frac{5}{6} + \frac{7}{8} = \frac{5 \cdot 4}{6 \cdot 4} + \frac{7 \cdot 3}{8 \cdot 3}$ The LCM of the denominator is _____.

$= \boxed{} + \boxed{}$ Write equivalent _____ using the common _____.

$= \boxed{} = \boxed{}$ Add.

LESSON 3-8 CONTINUED

Add or subtract. Write each answer in simplest form.

B. $\frac{2}{3} - \frac{3}{4}$

$\frac{2}{3} - \frac{3}{4} = \frac{2 \cdot 4}{3 \cdot 4} - \frac{3 \cdot 3}{4 \cdot 3}$ Multiply the _____.

$= \boxed{} - \boxed{}$ Write _____ fractions using the _____ denominator.

$= \boxed{}$ Subtract.

Example 3

In one Earth year, Jupiter completes about $\frac{1}{12}$ of its orbit around the Sun, while Mars completes about $\frac{1}{2}$ of its orbit. How much more of its orbit does Mars complete than Jupiter?

$\frac{1}{2} - \frac{1}{12} = \boxed{} - \frac{1}{12}$ The LCM of the denominators is _____.

$= \boxed{} - \frac{1}{12}$ Write _____ fractions using the common _____.

$= \boxed{}$ Subtract.

Mars completes _____ more of its orbit than Jupiter does.

Try This

1. Add. Write the answer in simplest form.

   $\frac{5}{6} + \frac{1}{6}$

2. Subtract. Write the answer in simplest form.

   $\frac{2}{5} - \frac{1}{2}$

LESSON 3-9: Adding and Subtracting Mixed Numbers

Lesson Objectives
Add and subtract mixed numbers

Additional Examples

Example 1
Kevin is $48\frac{3}{8}$ inches tall. His brother Keith is $5\frac{5}{8}$ inches taller. How tall is Keith?

$48\frac{3}{8} + 5\frac{5}{8} = \boxed{} + \dfrac{\boxed{}}{8}$ $\boxed{}$ the integers and $\boxed{}$ the fractions.

$= \boxed{} + \boxed{}$ Rewrite $\frac{8}{8}$ as $\boxed{}$.

Kevin is $\boxed{}$ inches tall. Add.

Example 2
Add. Write each answer in simplest form.

A. $9\frac{2}{3} + 12\frac{2}{3}$

$9\frac{2}{3} + 12\frac{2}{3} = \boxed{} + \dfrac{\boxed{}}{3}$ Add the $\boxed{}$ and add the $\boxed{}$.

$= \boxed{} + \boxed{}$ Rewrite the improper fraction as a $\boxed{}$ number.

$= \boxed{}$ Add.

B. $5\frac{1}{8} + 3\frac{5}{6}$

$5\frac{1}{8} + 3\frac{5}{6} = 5\boxed{} + 3\boxed{}$ Find a $\boxed{}$ denominator.

$= \boxed{} + \boxed{}$ $\boxed{}$ the integers and $\boxed{}$ the fractions.

$= \boxed{}$ Add.

LESSON 3-9 CONTINUED

Example 3

Subtract. Write each answer in simplest form.

A. $4\frac{2}{3} - 2\frac{1}{3}$

$4\frac{2}{3} - 2\frac{1}{3} = \square + \frac{\square}{3}$ ⎯⎯⎯⎯⎯⎯⎯ integers and ⎯⎯⎯⎯⎯⎯⎯ fractions.

$= \square$ Add.

B. $12\frac{8}{9} - 8\frac{2}{3}$

$12\frac{8}{9} - 8\frac{2}{3} = 12\square - 6\square$ Find a common denominator.

$= \square + \frac{\square}{9}$ Subtract ⎯⎯⎯⎯⎯⎯ and subtract ⎯⎯⎯⎯⎯⎯.

$= \square$ Add.

Try This

1. Alvin weighs $72\frac{2}{3}$ lbs. His baby brother weighs $15\frac{1}{3}$ lbs. How much do they weigh together?

2. Add. Write the answer in simplest form.

$7\frac{4}{5} + 10\frac{2}{5}$

Holt Mathematics

LESSON 3-10: Multiplying Fractions and Mixed Numbers

Lesson Objectives

Multiply fractions and mixed numbers

Additional Examples

Example 1

In 2001, the car toll on the George Washington Bridge was $6.00. In 1995, the toll was $\frac{2}{3}$ of that toll. What was the toll in 1995?

6 · ☐ = ☐ + ☐ + ☐ + ☐ + ☐

= ☐

= ☐   Simplify.

= $ ☐   Write the fraction as a decimal.

The George Washington Bridge toll for a car was $ ☐ in 1995.

Example 2

Multiply. Write each answer in simplest form.

A. $-12 \cdot \frac{3}{4}$

$-12 \cdot \frac{3}{4}$ = ☐ · $\frac{3}{4}$ Write -12 as a _____.

$= -\dfrac{\cancel{12}^{3} \cdot 3}{1 \cdot \cancel{4}_{1}}$ Simplify.

= ☐ = ☐   _____ numerators.

☐   _____ denominators.

B. $\frac{1}{3} \cdot \frac{3}{8}$ Simplify.

$\dfrac{1}{3} \cdot \dfrac{3}{8} = \dfrac{1 \cdot \cancel{3}^{1}}{_{1}\cancel{3} \cdot 8}$ _____ numerators.

= ☐   _____ denominators.

LESSON 3-10 CONTINUED

Example 3

Multiply. Write each answer in simplest form.

A. $\frac{2}{5} \cdot 1\frac{2}{3}$

$\frac{2}{5} \cdot 1\frac{2}{3} = \frac{2}{5} \cdot $ Write the _____ number as an improper _____.

$= \frac{2}{\cancel{5}_1} \cdot \frac{\cancel{5}^1}{3}$ Simplify.

$= $ _____ numerators.

_____ denominators.

B. $4\frac{1}{5} \cdot 2\frac{1}{7}$

$4\frac{1}{5} \cdot 2\frac{1}{7} = \cdot $ Write the mixed _____ as _____ fractions.

$= \frac{{}^3\cancel{21} \cdot \cancel{15}^3}{{}_1\cancel{5} \cdot \cancel{7}_1}$ Simplify.

$= $ or Multiply _____.

Multiply _____.

Try This

1. In 2002, the fee to park in a parking garage was $4. In 2000, the fee was $\frac{3}{4}$ of the fee in 2002. What was the fee in 2000?

2. Multiply. Write the answer in simplest form.

$-\frac{3}{7} \cdot \frac{1}{8}$

LESSON 3-11: Dividing Fractions and Mixed Numbers

Lesson Objectives
Divide fractions and mixed numbers

Vocabulary
reciprocal (p. 200) _____

Additional Examples

Example 1

Divide. Write each answer in simplest form.

A. $\dfrac{3}{7} \div \dfrac{2}{5}$

$\dfrac{3}{7} \div \dfrac{2}{5} = \dfrac{3}{7} \cdot \boxed{}$ Multiply by the _____ of $\dfrac{2}{5}$.

$= \dfrac{3 \cdot 5}{7 \cdot 2}$

$= \boxed{}$ or _____

B. $\dfrac{3}{8} \div 12$

$\dfrac{3}{8} \div 12 = \dfrac{3}{8} \cdot \boxed{}$

$= \dfrac{\overset{1}{3}}{8} \cdot \dfrac{1}{\underset{4}{12}}$ Multiply by the _____ of 12.

$= \boxed{}$ Simplify.

LESSON 3-11 CONTINUED

Example 2

Divide. Write each answer in simplest form.

A. $5\frac{2}{3} \div 1\frac{1}{4}$

$5\frac{2}{3} \div 1\frac{1}{4} = \div $ Write mixed _____ as improper _____.

$= \frac{17}{3} \cdot $ Multiply by the _____ of $\frac{5}{4}$.

$= $ or

B. $\frac{3}{4} \div 2\frac{1}{2}$

$\frac{3}{4} \div 2\frac{1}{2} = \frac{3}{4} \div $ Write $2\frac{1}{2}$ as an _____ fraction.

$= \frac{3}{4} \cdot $

$= \dfrac{3 \cdot \overset{1}{\cancel{2}}}{\underset{2}{\cancel{4}} \cdot 5}$ Multiply by the _____ of $\frac{5}{2}$.

$= $ Simplify.

C. $5\frac{5}{8} \div \frac{5}{9}$

$5\frac{5}{8} \div \frac{5}{9} = \div \frac{5}{9}$ Write $5\frac{5}{8}$ as an _____ fraction.

$= \frac{45}{8} \cdot $ Multiply by the _____ of $\frac{5}{9}$.

$= \dfrac{\overset{9}{\cancel{45}} \cdot 9}{8 \cdot \underset{1}{\cancel{5}}}$

$= $ or

LESSON 3-11 CONTINUED

Example 3

The life span of a golden dollar coin is 30 years, while paper currency lasts an average of $1\frac{1}{2}$ years. How many times longer will the golden dollar stay in circulation?

$30 \div 1\frac{1}{2} = \frac{30}{1} \div \frac{3}{2}$ Write the number as an ▢ fraction.

$= \frac{30}{1} \cdot$ ▢   Multiply by the ▢ of $\frac{3}{2}$.

$= \frac{{}^{10}\cancel{30} \cdot 2}{1 \cdot \cancel{3}_1}$ Simplify.

$=$ ▢ or ▢

The golden dollar will stay in circulation about ▢ times longer than paper currency.

Try This

1. Divide. Write the answer in simplest form.

   $\frac{3}{5} \div \frac{1}{2}$

2. Divide. Write the answer in simplest form.

   $\frac{3}{5} \div 1\frac{2}{5}$

3. The average life of a queen ant is approximately 3 years. The life span of a worker ant is $\frac{3}{7}$ year. How many times longer will the queen ant live?

LESSON 3-12: Solving Equations Containing Fractions

Lesson Objectives

Solve one-step equations that contain fractions

Additional Examples

Example 1

Solve. Write each answer in simplest form.

A. $x - \frac{3}{7} = \frac{5}{7}$

$$x - \frac{3}{7} = \frac{5}{7}$$

$$x - \frac{3}{7} + \boxed{} = \frac{5}{7} + \boxed{} \qquad \text{Add to } \boxed{} \; x.$$

$$x = \boxed{} = \boxed{} \qquad \text{Simplify.}$$

B. $\frac{5}{12} + t = \frac{3}{8}$

$$\frac{5}{12} + t = \frac{3}{8}$$

$$\frac{5}{12} + t - \boxed{} = \frac{3}{8} - \boxed{} \qquad \text{Subtract to isolate } \boxed{}.$$

$$t = \boxed{} - \boxed{} \qquad \text{Find a common } \boxed{}.$$

$$t = \boxed{} \qquad \text{Subtract.}$$

Example 2

Solve. Write each answer in simplest terms.

A. $\frac{3}{8}x = \frac{1}{4}$

$$\frac{3}{8}x = \frac{1}{4}$$

$$\frac{3}{8}x \cdot \frac{8}{3} = \frac{1}{\cancel{4}_1} \cdot \frac{\cancel{8}^2}{3} \qquad \text{Multiply by the } \boxed{} \text{ of } \frac{3}{8}.$$

$$x = \boxed{} \qquad \text{Then simplify.}$$

LESSON 3-12 CONTINUED

B. $4x = \frac{8}{9}$

$$4x = \frac{8}{9}$$

$$4x \cdot \frac{1}{4} = \frac{\overset{2}{\cancel{8}}}{9} \cdot \frac{1}{\cancel{4}_1}$$ Multiply by the reciprocal of ____.

$$x = $$ Then simplify.

Example 3

The amount of copper in brass is $\frac{3}{4}$ of the total weight. If a sample contains $4\frac{1}{5}$ ounces of copper, what is the total weight of the sample?

Let w represent the total weight of the sample.

$$\frac{3}{4}w = 4\frac{1}{5}$$ Write an equation.

$$\frac{3}{4}w \cdot = 4\frac{1}{5} \cdot $$ Multiply by the ____ of $\frac{3}{4}$.

$$w = \frac{\overset{7}{\cancel{21}}}{5} \cdot \frac{4}{\underset{1}{\cancel{3}}}$$ Write $4\frac{1}{5}$ as an ____ fraction.

$$w = \text{ or } $$ Then simplify.

The sample weighs ____ ounces.

Try This

1. Solve. Write the answer in simplest form.

$$x - \frac{3}{8} = \frac{7}{8}$$

Chapter Review

3-1 Estimate with Decimals

Estimate by rounding to the nearest integer.

1. $26.79 + 44.3$
2. $11.419 - 6.94$
3. $-13.581 + 7.8$
4. $8.66 + (-19.3)$

Use compatible numbers to estimate.

5. $42.67 \cdot 9.14$
6. $73.4 \div 7.96$
7. $-6.83 \cdot 38.721$
8. $148.39 \div (-6.23)$

9. Leah is saving for an iPod that costs $189.55. She already has $82.38 and saves $12.25 each week. About how many weeks will it take Leah to save enough money to buy the iPod?

3-2 Adding and Subtracting Decimals

Add or subtract. Estimate to check whether each answer is reasonable.

10. $-3.817 + 4.2$
11. $7.624 - 18.34$
12. $-4.77 - 12.053$
13. $30.62 - (-9.18)$

14. The table shows the top 5 movies of all time and the amount each made in profit.

Movie	$ (million)
Titanic (1997)	600.8
Star Wars (1977)	460.9
Shrek 2 (2004)	436.5
E. T. The Extra-Terrestrial (1982)	434.9
Star Wars Episode I – The Phantom Menace (1999)	431.1

a) How much more money did Titanic (1997) make in profit than Star Wars (1977)?

b) How much money have the top 3 movies made in profit altogether?

CHAPTER 3 REVIEW CONTINUED

3-3 Multiplying Decimals

Multiply. Estimate to check whether each answer is reasonable.

15. −8.7 · 3.12   **16.** 1.89 · 0.07   **17.** −5.21 · (−43.6)   **18.** 0.58 · (−3.1)

19. Fresh ground beef is on sale for $4.90 per pound. How much will it cost to buy 8.30 pounds of ground beef?

3-4 Dividing Decimals by Integers

Divide. Estimate to check whether each answer is reasonable.

20. −37.22 ÷ 4   **21.** 0.86 ÷ (−8)   **22.** 22.92 ÷ 12   **23.** −61.8 ÷ 15

Simplify each expression. Justify each step using the Commutative, Associative, or Distributive Property when necessary.

24. 23.8 ÷ 5 · 7 + 8.9

25. −48.9 + 11.64 ÷ (21.3 − 16.3)^2

26. A plane burned 173.82 gallons of gasoline for a flight that lasted 12 hours. How many gallons of gasoline were burned per hour?

3-5 Dividing Decimals and Integers by Decimals

Divide. Estimate to check whether each answer is reasonable.

27. 48 ÷ 3.2   **28.** −13.6 ÷ 1.6   **29.** 72 ÷ (−3.2)   **30.** −154 ÷ (−0.35)

Simplify each expression. Justify each step using the Commutative, Associative, or Distributive Property when necessary.

31. (10.8 ÷ 0.6) · (−6.4 + 9.3)

32. 20 · 15 · (−5.2 ÷ 2.6) · 3

CHAPTER 3 REVIEW CONTINUED

3-6 Solving Equations Containing Decimals

Solve.

33. $t - 0.94 = 18.5$

34. $w + 28.52 = -16.03$

35. $0.27x = -8.1$

36. $\dfrac{k}{-0.36} = 8.5$

37. Shea bought CDs for $11.65 each. She spent a total of $163.10. How many CDs did Shea buy?

3-7 Estimate With Fractions

Estimate each sum, difference, product, or quotient.

38. $4\dfrac{3}{5} - 2\dfrac{5}{12}$

39. $15\dfrac{1}{3} + 2\dfrac{8}{9} - 9\dfrac{5}{11}$

40. $4\dfrac{1}{5} \cdot 7\dfrac{7}{9}$

41. $23\dfrac{6}{7} \div (-1\dfrac{1}{12})$

3-8 Adding and Subtracting Fractions

Find each sum or difference. Write your answer in simplest form.

42. $-\dfrac{1}{4} + \dfrac{3}{8} + \dfrac{5}{12}$

43. $\dfrac{5}{6} - \dfrac{3}{15} - \dfrac{2}{5}$

44. Kevin ran for $\dfrac{3}{4}$ hour, swam for $\dfrac{1}{2}$ hour, and bicycled for $\dfrac{5}{6}$ hour. How long did Kevin run, swim, and bike altogether?

3-9 Adding and Subtracting Mixed Numbers

Compare. Write <, >, or =.

45. $10\dfrac{1}{4} - 4\dfrac{3}{4}$ ■ $12\dfrac{3}{5} - 6\dfrac{9}{10}$

46. $16\dfrac{7}{8} + 4\dfrac{3}{4}$ ■ $25\dfrac{1}{8} - 3\dfrac{1}{2}$

47. Maggie hiked $7\dfrac{5}{8}$ miles at Turkey Run State Park, and $12\dfrac{3}{10}$ miles at Starved Rock State Park. How many more miles did Maggie hike at Starved Rock than Turkey State Park?

CHAPTER 3 REVIEW CONTINUED

3-10 Multiplying Fractions and Mixed Numbers

Complete each multiplication sentence.

48. $\dfrac{2}{3} \cdot \dfrac{\blacksquare}{9} = \dfrac{10}{27}$ 49. $\dfrac{7}{8} \cdot \dfrac{2}{\blacksquare} = \dfrac{7}{12}$ 50. $\dfrac{9}{\blacksquare} \cdot \dfrac{4}{5} = \dfrac{9}{25}$

51. Caitlin drove at a speed of 65 mph for $6\dfrac{3}{4}$ hours. How many miles did she travel?

3-11 Dividing Fractions and Mixed Numbers

Evaluate. Write each answer in simplest form.

52. $\dfrac{4}{5} \div 6\dfrac{2}{3}$

53. $2\dfrac{7}{8} \div \left(\dfrac{1}{3} + \dfrac{3}{4}\right)$

54. $\dfrac{5}{6} \cdot \dfrac{2}{3} \div 2\dfrac{2}{9}$

55. $3\dfrac{1}{3} \div \left(\dfrac{3}{5} \cdot \dfrac{1}{2}\right)$

56. Tommy and his three brothers worked for $177\dfrac{3}{5}$ hours. What was the average number of hours each brother worked?

3-12 Solving Equations Containing Fractions

Solve. Write each answer in simplest form.

57. $\dfrac{5}{7}x = \dfrac{5}{8}$

58. $w + \dfrac{3}{4} = \dfrac{11}{12}$

59. $-\dfrac{17}{55} + d = \dfrac{28}{55}$

60. $2\dfrac{5}{18}p = 5\dfrac{4}{9}$

61. Mark spends $\dfrac{1}{3}$ of his day sleeping and $\dfrac{1}{4}$ of his day at school. What fraction of his day is spent doing things besides sleeping and going to school?

CHAPTER 3 Big Ideas

Answer these questions to summarize the important concepts from Chapter 3 in your own words.

1. Explain how to round a decimal to the nearest integer.

2. Explain how to multiply decimals.

3. Explain how to divide an integer by a decimal.

4. Explain the guidelines for rounding fractions.

5. Explain the difference between adding or subtracting fractions with unlike denominators, and multiplying fractions with unlike denominators.

For more review of Chapter 3:

- Complete the Chapter 3 Study Guide and Review on pages 212–214 of your textbook.
- Complete the Ready to Go On quizzes on pages 178 and 208 of your textbook.

LESSON 4-1

The Coordinate Plane

Lesson Objectives

Plot and identify ordered pairs on a coordinate plane

Vocabulary

coordinate plane (p. 224)

x-axis (p. 224)

y-axis (p. 224)

origin (p. 224)

quadrant (p. 224)

ordered pair (p. 224)

Additional Examples

Example 1

Identify the quadrant that contains each point.

A. S

   S lies in Quadrant _____ .

B. T

   T lies in Quadrant _____ .

C. W

   W lies on the _____ between

   Quadrants _____ and _____ .

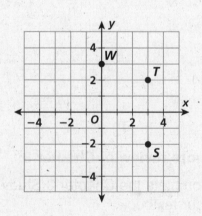

LESSON 4-1 CONTINUED

Example 2

Plot each point on a coordinate plane.

A. D (3, 3)

Start at the _____. Move ____ units right and ____ units up.

B. E (−2, −3)

Start at the _____. Move 2 units _____ and 3 units _____.

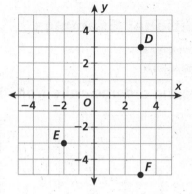

Example 3

Give the coordinates of each point.

A. X

Start at the _____. Point X is ____ units left and ____ units up.

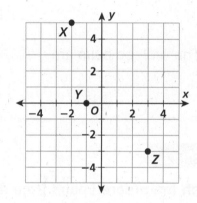

Try This

1. Identify the quadrant that contains the point.

   X

   _____

2. Plot the point on the coordinate plane.

   E (−2, 3)

3. Give the coordinates of the point.

   L

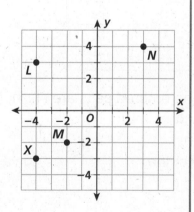

LESSON 4-2 Tables and Graphs

Lesson Objectives
Identify and graph ordered pairs from a table of values

Additional Examples

Example 1

Write the ordered pairs from the table.

x	y
-2	-1
4	7
0	9
5	-3

(x, y)
(,)
(,)
(,)
(,)

The ordered pairs are _____, _____, _____, and _____.

Example 2

Graph the ordered pairs from the table.

x	y
1	4
-1	-2
3	2
-5	0

The ordered pairs are (1, 4), (−1, −2), (3, 2), and (−5, 0).

Plot the _____ on a coordinate plane.

Holt Mathematics

LESSON 4-2 CONTINUED

Example 3

What appears to be the relationship between the number of pounds and the cost of food shown in the table below?

Number of Pounds	1	2	3	4
Cost ($)	1.85	3.70	5.55	7.40

Write the ordered pairs from the table.

Number of Pounds	1	2	3	4
Cost ($)	1.85	3.70	5.55	7.40

(x, y)	(1, 1.85)	(2, 3.70)	(3, 5.55)	(4, 7.40)

The ordered pairs are _____ , _____ , _____ ,

_____ .

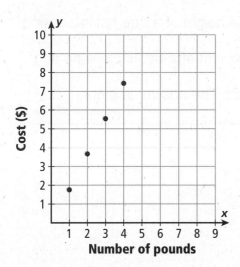

Lesson 4-3: Interpreting Graphs

Lesson Objectives
Relate graphs to situations

Additional Examples

Example 1

The height of a tree increases over time, but not at a constant rate. Which graph bests shows this?

a.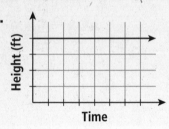

b.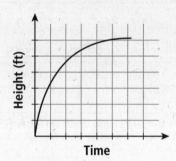

As the tree grows, its height _____ and then reaches its _____ height. Graph ____ shows the height of a tree not increasing but remaining constant. Graph ____ shows the height of a tree increasing at a constant rate without reaching a maximum height. The answer is graph ____.

LESSON 4-3 CONTINUED

Example 2

Jarod parked his car in the supermarket parking lot and walked 40 ft into the store to the customer service counter, where he waited in line to pay his electric bill. Jarod then walked 60 ft to the back of the store to get 2 gallons of milk and walked 50 ft to the checkout near the front of the store to pay for them. After waiting his turn and paying for the milk, he walked 50 ft back to his car. Sketch a graph to show Jarod's distance from his car over time.

1. **Understand the Problem**

   The answer is the _____ showing _____ Jarod traveled.

   List the **important information**:

   - Jarod walked to the _____

   - Jarod _____ in line.

   - Jarod walked to the _____ of the store.

   - Jarod walked to the _____.

   - Jarod _____ in line.

   - Jarod went back to his _____.

2. **Make a Plan**

   Sketch a graph of the situation.

3. **Make a Plan**

   The distance _____ as Jarod walks to the customer service counter.

   The distance _____ when Jarod waits in line.

   The distance _____ as Jarod walks to the bar of the store.

LESSON 4-3 CONTINUED

The distance _____ as Jarod walks to the checkout.

The distance _____ when Jarod waits in line.

The distance _____ as Jarod walks back to his car.

4. Look Back

The graph is reasonable because it pictures someone walking away, standing still, walking away, walking back, standing in line, and walking back.

Try This

1. The dimensions of the basketball court have changed over the years. However, the height of the basket has not changed. Which graph bests shows this?

   a.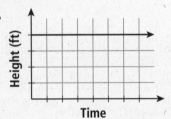

   b.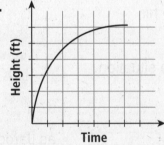

LESSON 4-4: Functions, Tables, and Graphs

Lesson Objectives

Use function tables to generate and graph ordered pairs

Vocabulary

function (p. 238)

Additional Examples

Example 1

Find the output for the input.

$y = 8x + 5$

Input	Rule	Output
x	8x + 5	y
−4	8( ) + 5	
−2	8( ) + 5	
1	8( ) + 5	

Substitute _____ for *x* and simplify.

Substitute _____ for *x* and simplify.

Substitute _____ for *x* and simplify.

LESSON 4-4 CONTINUED

Example 2

Make a function table and graph the resulting ordered pairs.

$y = 3x - 4$

Input	Rule	Output	Ordered Pair
x	3x − 4	y	(x, y)
−2	3( ) − 4		
−1	3( ) − 4		
0	3( ) − 4		
1	3( ) − 4		
2	3( ) − 4		

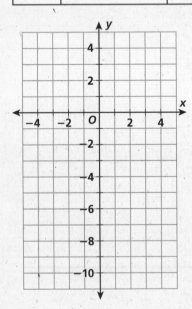

LESSON 4-5: Find a Pattern in Sequences

Lesson Objectives

Find patterns to complete sequences using function tables

Vocabulary

sequence (p. 242)

arithmetic sequence (p. 242)

geometric sequence (p. 242)

Additional Examples

Example 1

Tell whether the sequence of y-values is arithmetic or geometric. Then find y when n = 5.

A.

n	1	2	3	4	5
y	−1	−4	−16	−64	

In the sequence, −1, −4, −16, −64, each number is multiplied by ____.

−64 · ____ = ____   Multiply the fourth number by ____.

The sequence is _____. When $n = 5$, $y = $ ____.

B.

n	1	2	3	4	5
y	51	46	41	36	

In the sequence, 51, 46, 41, 36, ____ is added to each number.

36 + ____ = ____   Add ____ to the fourth number.

The sequence is _____. When $n = 5$, $y = $ ____.

LESSON 4-5 CONTINUED

Example 2

Write a function that describes the sequence 3, 6, 9, 12, ...

Make a function table.

n	Rule	y
1	1 · ⬚	3
2	2 · ⬚	6
3	3 · ⬚	9

The function ⬚ describes this sequence.

Example 3

Holli keeps a list showing her cumulative earnings for walking her neighbor's dog. She recorded $1.25 for the first time she walked the dog, $2.50 for the second time, $3.75 the third time, and $5.00 the fourth time. Write a function that describes the sequence, and then use the function to predict her earnings after 9 walks.

Write the amounts to money she collected: $1.25, $2.50, $3.75, $5.00, ...

Make a function table.

n	Rule	y
1	1 · ⬚	1.25
2	2 · ⬚	2.50
3	3 · ⬚	3.75
4	4 · ⬚	5.00

Multiply n by ⬚.

⬚   Write the function.

After 9 walks corresponds to $n =$ ⬚. When $n =$ ⬚,

$y = 1.25 \cdot$ ⬚ $=$ ⬚.

Holli earns ⬚ after 9 walks.

LESSON 4-6: Graphing Linear Functions

Lesson Objectives

Identify and graph linear equations

Vocabulary

linear equation (p. 248)

linear function (p. 248)

Additional Examples

Example 1

Graph the linear function $y = 4x - 1$.

Input	Rule	Output	Ordered Pair
x	4x − 1	y	(x, y)
0	4( ) − 1		
1	4( ) − 1		
−1	4( ) − 1		

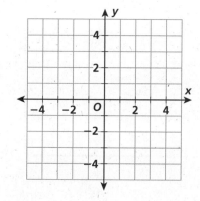

Place each ordered _____ on the coordinate grid and then connect the points with a _____.

LESSON 4-6 CONTINUED

Example 2

The fastest-moving tectonic plates on Earth move apart at a rate of 15 centimeters per year. Scientists began studying two parts of these plates when they were 30 centimeters apart. Write a linear function that describes the movement of the plates over time. Then make a graph to show the movement over 4 years.

The function is _____, where x is the number of years and y is the spread in centimeters.

Input	Rule	Output
x	15x + 30	y
0	15( ) + 30	
2	15( ) + 30	
4	15( ) + 30	

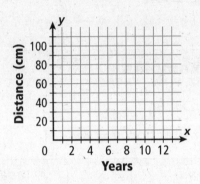

Try This

1. Graph the linear function y = 3x + 1.

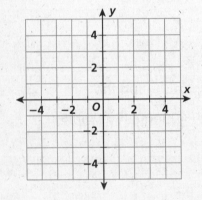

Copyright © by Holt, Rinehart and Winston.
All rights reserved.

Holt Mathematics

Chapter Review

CHAPTER 4

4-1 The Coordinate Plane

Identify the quadrant that contains each point.

1. (7, 3)
2. (−5, −1)
3. (−3, 0)
4. (2, −2)

Plot each ordered pair on a coordinate plane.

5. (3, −5)
6. (0, 4)
7. (−1, 6)
8. (4, 2)

4-2 Tables and Graphs

Write and graph the ordered pairs from each table.

9.

x	y
−2	2
−1	2
0	2
1	2

10.

x	y
−2	−3
−1	−1
1	3
2	5

11. The table shows the total cost of buying gasoline for a car. Graph the data to find the cost of buying 12 gallons of gasoline?

Number of Gallons	Price
2	$2.50
4	$5.00
6	$7.50
8	$10.00

CHAPTER 4 REVIEW CONTINUED

4-3 Interpreting Graphs

12. An airplane increases in altitude from take-off until it reaches its cruising altitude. It flies at the cruising altitude until it begins to descend for landing. Which graph best shows the story?

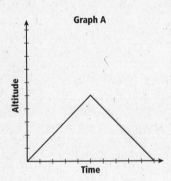
Graph A

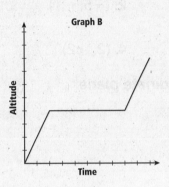
Graph B

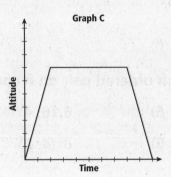
Graph C

13. Nolan rode his bike to school. After school he rode his bike to a friend's house. Then he rode his bike home. Sketch a graph to show the distance Nolan traveled.

4-4 Functions, Tables, and Graphs

Make a function table, and graph the resulting ordered pairs.

14. $y = 5x - 7$

Input	Rule	Output	Ordered Pair
x	5x − 7	y	(x, y)
−1			
0			
1			
2			

CHAPTER 4 REVIEW CONTINUED

4-5 Find a Pattern in Sequences

Tell whether the sequence of y-values is arithmetic or geometric. Then find y when n = 4.

15.
n	1	2	3	4
y	−2	−5	−8	

16.
n	1	2	3	4
y	20	10	5	

Find a function that describes each sequence. Use the function to find the eighth term in the sequence.

17. −3, −6, −9, −12, . . .

18. 8, 9, 10, 11, . . .

4-6 Graphing Linear Functions

Graph each linear function.

19. $y = 3x - 2$

20. $y = -2x + 4$

21. A car is traveling 60 miles per hour. Write a linear function that describes the distance the car travels over time. Then make a graph to show the distance the car travels in 9 hours.

CHAPTER 4 Big Ideas

Answer these questions to summarize the important concepts from Chapter 4 in your own words.

1. Explain how to plot the ordered pair $(4, -7)$.

2. Explain how to find the output value for the function $y = 6x + 2$ for $x = -5$.

3. Explain the difference between an arithmetic sequence and a geometric sequence.

4. Explain how to graph the linear function $y = 3x - 2$.

5. In the function $y = 6x + 8$, which is the dependent variable and which is the independent variable? Explain.

For more review of Chapter 4:

- Complete the Chapter 4 Study Guide and Review on pages 258–260 of your textbook.
- Complete the Ready to Go On quizzes on pages 236 and 252 of your textbook.

Lesson 5-1: Ratios

Lesson Objectives
Identify, write, and compare ratios

Vocabulary
ratio (p. 270)

Additional Examples

Example 1

Twenty students are asked to choose their favorite music category. Eight chose pop, seven chose hip hop, and five chose rock. Write each ratio in all three forms.

A. rock to hip hop

There were ____ students that chose rock and ____ students that chose hip hop.

The ratio of rock to hip hop is ____ to ____, which can be written as follows:

B. hip hop to pop

There were ____ students that chose hip hop and ____ students that chose pop.

The ratio of hip hop to pop is ____ to ____, which can be written as follows:

C. rock to pop and hip hop

There were ____ students that chose rock and ____ + ____ = ____ students that chose hip hop and pop.

The ratio of rock to hip hop and pop is ____ to ____, which can be written as follows:

LESSON 5-1 CONTINUED

Example 2

On average, most people can read about 600 words in 3 minutes. Write the ratio of words to minutes in all three forms. Write your answers in simplest form.

$\dfrac{\text{words}}{\text{minutes}} = \dfrac{\boxed{} \text{ words}}{\boxed{} \text{ minutes}}$ Write the ratio as a fraction.

$\dfrac{\text{words}}{\text{minutes}} = \dfrac{600 \div \boxed{}}{3 \div \boxed{}}$ Simplify.

$\dfrac{\text{words}}{\text{minutes}} = \dfrac{\boxed{} \text{ words}}{1 \text{ minute}}$ For every minute, _____ words are typed.

The ratios in simplest form are: _____.

Example 3

Honey lemon cough drops come in packages of 30 drops per 10-ounce bag. Cherry cough drops come in packages of 24 drops per 6-ounce bag. Tell which package has the greater ratio of drops to ounces.

Honey: $\dfrac{\text{drops}}{\text{ounces}} = \dfrac{\boxed{} \text{ drops}}{\boxed{} \text{ ounces}} = \dfrac{\boxed{}}{\boxed{}}$ Write the ratios fractions with common denominators.

Cherry: $\dfrac{\text{drops}}{\text{ounces}} = \dfrac{\boxed{} \text{ drops}}{\boxed{} \text{ ounces}} = \dfrac{\boxed{}}{\boxed{}}$

Because _____ > _____ and the denominators are the _____, the bag of _____ has the greater ratio of drops to ounces.

LESSON 5-2: Rates

Lesson Objectives

Find unit and compare unit rates, such as average speed and unit price

Vocabulary

rate (p. 274)

unit rate (p. 274)

Additional Examples

Example 1

Find each rate.

A. A Ferris wheel revolves 35 times in 105 minutes. How many minutes does 1 revolution take?

☐ minutes / ☐ revolutions   Write a rate that compares minutes and revolutions.

☐ minutes ÷ ☐ / ☐ revolutions ÷ ☐   Divide the numerator and denominator by ☐.

☐ minutes / 1 revolution   Simplify.

The Ferris wheel takes ☐ minutes for 1 revolution.

LESSON 5-2 CONTINUED

B. Sue walks 6 yards and passes 24 security lights set along the sidewalk. How many security lights does she pass in 1 yard?

$\dfrac{\boxed{} \text{ lights}}{\boxed{} \text{ yards}}$ Write a rate that compares lights and yards.

$\dfrac{\boxed{} \text{ lights} \div \boxed{}}{\boxed{} \text{ yards} \div \boxed{}}$ Divide the numerator and denominator by $\boxed{}$.

$\dfrac{\boxed{} \text{ lights}}{1 \text{ yard}}$ Simplify.

Sue walks past $\boxed{}$ security lights in 1 yard.

Example 2

Danielle is cycling 68 miles as a fundraising commitment. She wants to complete her ride in 4 hours. What should be her average speed in miles per hour?

$\dfrac{\boxed{} \text{ miles}}{\boxed{} \text{ hours}}$ Write the rate as a fraction.

$\dfrac{\boxed{} \text{ miles} \div \boxed{}}{\boxed{} \text{ hours} \div \boxed{}} = \dfrac{\boxed{} \text{ miles}}{1 \text{ hour}}$ Divide the numerator and denominator by $\boxed{}$.

Her average speed should be $\boxed{}$ miles per hour.

LESSON 5-2 CONTINUED

Example 3

A 12-ounce sports drink costs $0.99, and a 16-ounce drink costs $1.19. Which size is the best buy?

Divide the _____ by the number of _____ (oz) to find the unit price of each size.

$\dfrac{\$\;\rule{1cm}{0.15mm}}{\rule{1cm}{0.15mm}\;\text{oz}} \approx \dfrac{\$\;\rule{1cm}{0.15mm}}{\text{oz}}$ $\dfrac{\$\;\rule{1cm}{0.15mm}}{\rule{1cm}{0.15mm}\;\text{oz}} \approx \dfrac{\$\;\rule{1cm}{0.15mm}}{\text{oz}}$

Since $ _____ < $ _____ , the _____ -oz sports drink is the best buy.

Try This

1. Find the rate.

A car gets 189 miles with 9 gallons of gas. How many miles does the car get in 1 gallon of gas?

2. Danielle is walking 18 miles for charity. She wants to complete her walk in 3 hours. What should be her average speed in miles per hour?

3. A 28-ounce box of cereal costs $2.99, and a 32-ounce box of cereal costs $3.19. Which size is the best buy?

LESSON 5-3

Slope and Rates of Change

Lesson Objectives

Determine the slope of a line and recognize constant and variable rates of change.

Vocabulary

slope (p. 278) _____

_____

Additional Examples

Example 1

Tell whether the slope is positive or negative. Then find the slope.

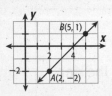

The line points _____.

The slope is _____.

slope = $\frac{\text{rise}}{\text{run}}$

= ——

The rise is ____. The run is ____.

= ☐

LESSON 5-3 CONTINUED

Example 2

Use the slope $\frac{-2}{1}$ and the point (1, −1) to graph the line.

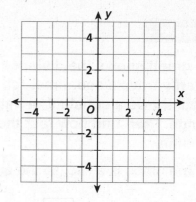

$\frac{\text{rise}}{\text{run}}$ = ——— or ———

From the point _____, move _____ units down and _____ unit right, or move _____ units up and _____ unit left. Mark the point where you end up, and draw a line through the two points.

Example 3

Tell whether the graph shows a constant or variable rate of change.

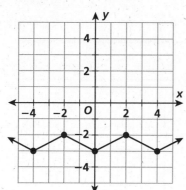

The graph is _____, so the rate of change is _____.

LESSON 5-3 CONTINUED

Example 4

The graph shows the distance a monarch butterfly travels over time. Tell whether the graph shows a constant or variable rate of change. Then find how fast the butterfly is traveling.

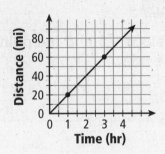

The graph is a line, so the butterfly is traveling at a _____ rate of speed.

The amount of _____ is the rise, and the amount of _____ is the run. You can find the speed by finding the _____.

slope = ▭/▭ = ▭ miles / hour

The butterfly travels at a rate of ▭ miles per hour.

Try This

1. Tell whether the slope is positive or negative. Then find the slope.

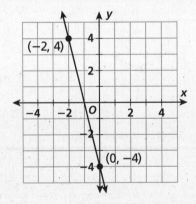

LESSON 5-4 Identifying and Writing Proportions

Lesson Objectives
Find equivalent ratios and identify proportions

Vocabulary
equivalent ratios (p. 283)

proportion (p. 283)

Additional Examples

Example 1

Determine whether the ratios are proportional.

A. $\frac{24}{51}, \frac{72}{128}$

$\frac{24 \div 3}{51 \div 3} =$ Simplify $\frac{24}{51}$.

$\frac{72 \div 8}{128 \div 8} =$ Simplify $\frac{72}{128}$.

Since $\neq$, the ratios   proportional.

B. $\frac{150}{105}, \frac{90}{63}$

$\frac{150 \div}{105 \div} = \frac{10}{7}$ Simplify $\frac{150}{105}$.

$\frac{90 \div 9}{63 \div 9} =$ Simplify $\frac{90}{63}$.

Since $\frac{10}{7}$, the ratios   proportional.

LESSON 5-4 CONTINUED

Example 2

Directions for making 12 servings of rice call for 3 cups of rice and 6 cups of water. For 40 servings, the directions call for 10 cups of rice and 19 cups of water. Determine whether the ratios of rice to water are proportional for both servings of rice.

Write the _____ of rice to water for 12 servings and for 40 servings.

Ratio of rice to water, 12 servings: ____   Write the ratio as a fraction.

Ratio of rice to water, 40 servings: ____   Write the ratio as a fraction.

$\frac{3}{6} = \frac{3 \cdot 19}{6 \cdot 19} =$ ____   Write the ratios with a _____ denominator, such as 114.

$\frac{10}{19} = \frac{10 \cdot 6}{19 \cdot 6} =$ ____

Since ____ ≠ ____ , the two ratios ____ proportional.

Example 3

Find an equivalent ratio. Then write the proportion.

A. $\frac{3}{5}$

$\frac{3}{5} = \frac{3 \cdot 2}{5 \cdot 2} =$ ____   Multiply both the _____ and _____ by any number such as 2.

____ = ____   Write a _____.

B. $\frac{28}{16}$

$\frac{28}{16} = \frac{28 \div 4}{16 \div 4} =$ ____   Divide both the _____ and _____ by any number such as 4.

____ = ____   Write a _____.

LESSON 5-5: Solving Proportions

Lesson Objectives
Solve proportions by using cross products

Vocabulary
cross product (p. 287) _____

Additional Examples

Example 1

Use cross products to solve the proportion.

$$\frac{9}{15} = \frac{m}{5}$$

☐ · m = ☐ · 5   The cross products are ☐.

m = ☐   Multiply.

☐ = ☐   Divide each side by ☐ to isolate the variable.

m = ☐

Example 2

PROBLEM SOLVING APPLICATION

If 3 volumes of Jennifer's encyclopedia take up 4 inches of space on her shelf, how much space will she need for all 26 volumes?

1. **Understand the Problem**
   Rewrite the question as a statement.

   • Find the space needed for ☐ volumes of the encyclopedia.

   List the important information:

   • ☐ volumes of the encyclopedia take up ☐ inches of space.

LESSON 5-5 CONTINUED

2. Make a Plan

Set up a proportion using the given information.

$\dfrac{3 \text{ volumes}}{4 \text{ inches}} = \dfrac{26 \text{ volumes}}{x}$ Let x be the unknown space.

3. Solve

$\dfrac{3}{4} = \dfrac{26}{x}$ Write the _____.

☐ · x = 4 · ☐   The cross products are _____.

☐ x = 104   Multiply.

☐ = ☐   Divide each side by ☐ to isolate the variable.

x = ☐

She needs _____ inches for all 26 volumes.

4. Look Back

$\dfrac{3}{4} \diagup \dfrac{26}{34\frac{2}{3}}$ $4 \cdot 26 = 104$

$3 \cdot 34\dfrac{2}{3} = 104$

The cross products are equal, so $34\dfrac{2}{3}$ is the answer.

Try This

1. Use cross products to solve the proportion.

$\dfrac{6}{7} = \dfrac{m}{14}$

LESSON 5-6 Customary Measurements

Lesson Objectives

Identify and convert customary units of measure

Additional Examples

Example 1

Choose the most appropriate customary unit for each measurement. Justify your answer.

A. the weight of a car

Tons—the weight of a car is similar to the weight of a buffalo.

B. the diameter of a soup can

Inches—the diameter of a soup can is similar to the length of a few paper clips.

C. the weight of a newborn baby

Pounds—the weight of a newborn baby is similar to the weight of more than a dozen apples.

Example 2

Convert 5,000 pounds to tons.

Write a proportion using a ratio of equivalent _____.

_____ → _____ = _____

1 · _____ = _____ · x

_____ = x

5,000 pounds is equal to _____ tons.

LESSON 5-6 CONTINUED

Example 3

Allison orders a 6-ounce grilled chicken sandwich. Dominic orders a quarter-pound grilled chicken sandwich. Which sandwich weighs more? Explain.

First convert one-quarter pound to ☐.

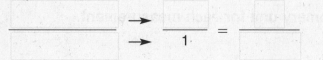 → ☐/1 = ☐/☐   Write a proportion using ☐ lb = ☐ oz.

16 · ☐ = ☐ · x

☐ = x

☐ sandwich weighs more. ☐ sandwich weighs only ☐ oz.

Try This

1. Choose the most appropriate customary unit for the measurement. Justify your answer.

    the capacity of a bathtub

2. Convert 255 feet to inches.

3. A pitcher had 2 gallons of lemonade. Amy drank 2 pints of the lemonade in the pitcher. How much lemonade remained in the pitcher?

LESSON 5-7: Similar Figures and Proportions

Lesson Objectives

Use ratios to determine if two figures are similar

Vocabulary

similar (p. 300)

corresponding sides (p. 300)

corresponding angles (p. 300)

Additional Examples

Example 1

Identify the corresponding sides in the pair of triangles. Then use ratios to determine whether the triangles are similar.

$\overline{AB}$ corresponds to ☐.

$\overline{BC}$ corresponds to ☐.

$\overline{AC}$ corresponds to ☐.

$\dfrac{AB}{DE} \stackrel{?}{=} \dfrac{BC}{EF} \stackrel{?}{=} \dfrac{AC}{DF}$ Write _____ using the corresponding sides.

☐ $\stackrel{?}{=}$ ☐ $\stackrel{?}{=}$ ☐   Substitute the lengths of the sides.

☐ = ☐ = ☐   Simplify each ratio.

Since the ratios of the corresponding sides are _____, the triangles _____ similar.

LESSON 5-7 CONTINUED

Example 2

Tell whether the figures are similar.

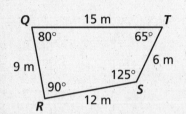

The _____ angles of the figures have _____ measures.

Write each set of sides as a ratio.

$\overline{MN}$ corresponds to $\overline{QR}$.

$\overline{NO}$ corresponds to $\overline{RS}$.

$\overline{OP}$ corresponds to $\overline{ST}$.

$\overline{MP}$ corresponds to $\overline{QT}$.

Determine whether the ratios of the lengths of the corresponding sides are _____.

$\dfrac{MN}{QR} \stackrel{?}{=} \dfrac{NO}{RS} \stackrel{?}{=} \dfrac{OP}{ST} \stackrel{?}{=} \dfrac{MP}{QT}$

Write each ratio using the _____ sides.

$\dfrac{6}{9} \stackrel{?}{=} \dfrac{8}{12} \stackrel{?}{=} \dfrac{4}{6} \stackrel{?}{=} \dfrac{10}{15}$

Substitute the lengths of the sides.

☐ = ☐ = ☐ = ☐   Simplify each _____.

Since the ratios of the corresponding sides are _____, the figures _____ similar.

LESSON 5-8 Using Similar Figures

Lesson Objectives

Use similar figures to find unknown lengths

Vocabulary

indirect measurement (p. 304)

Additional Examples

Example 1

Find the unknown length in the similar figures.

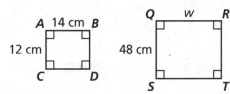

$\dfrac{AC}{QS} = \dfrac{AB}{QR}$ Write a proportion using corresponding _____.

_____ $= \dfrac{14}{w}$ Substitute the lengths of the _____.

$12 \cdot w =$ _____ $\cdot\, 14$ Find the _____ products.

$w =$ _____   Multiply.

_____ $=$ _____   Divide each side by _____ to isolate the variable.

$w =$ _____

QR is _____ centimeters.

LESSON 5-8 CONTINUED

Example 2

The inside triangle is similar in shape to the outside triangle. Find the length of the base of the inside triangle.

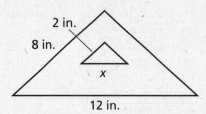

Let x = the _____ of the base of the inside triangle.

☐/☐ = ☐/x   Write a proportion using corresponding side lengths.

☐ · x = ☐ · ☐   Find the cross products.

x = ☐   Multiply.

☐x/☐ = ☐/☐   Divide each side by ☐ to isolate the variable.

x = ☐

The length of the base of the inside triangle is ☐ inches.

LESSON 5-8 CONTINUED

Example 3

City officials want to know the height of a traffic light. Estimate the height of the traffic light.

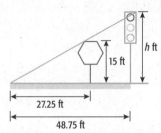

$\dfrac{\boxed{}}{\boxed{}} = \dfrac{\boxed{}}{\boxed{}}$ Write a _____.

$\dfrac{\boxed{}}{\boxed{}} = \dfrac{\boxed{}}{\boxed{}}$ Use _____ numbers to estimate.

$\dfrac{\boxed{}}{\boxed{}} = \boxed{}$ Simplify.

Multiply each side by _____ to isolate the variable.

$\boxed{} \cdot \dfrac{\boxed{}}{\boxed{}} = \boxed{} \cdot \boxed{}$

$h = \boxed{}$

The height of the traffic light is about _____ feet.

LESSON 5-9: Scale Drawings and Scale Models

Lesson Objectives

Understand ratios and proportions in scale drawings; use ratios and proportions with scale

Vocabulary

scale model (p. 308)

scale factor (p. 308)

scale (p. 308)

scale drawing (p. 308)

Additional Examples

Example 1

Identify the scale factor.

	Room	Blueprint
Length (in.)	144	18
Width (in.)	108	13.5

$\dfrac{\text{blueprint length}}{\text{room length}} =$ ☐   Write a _____ using one of the dimensions.

$=$ ☐   Simplify.

The scale factor is ☐.

LESSON 5-9 CONTINUED

Example 2

A photograph was enlarged and made into a poster. The poster is 20.5 inches by 36 inches. The scale factor is $\frac{5}{1}$. Find the size of the photograph.

Think: $\frac{\text{poster}}{\text{photo}} = \frac{5}{1}$

$\frac{36}{l} = \frac{5}{1}$ Write a _____ to find the length *l*.

$5l = $ ____   Find the _____ products.

$l = $ Divide.

$\frac{20.5}{w} = \frac{5}{1}$ Write a _____ to find the width *w*.

$w = $ Find the cross _____.

$w = $ Divide.

The photo is ____ in. long and ____ in. wide.

Example 3

On a road map, the distance between Pittsburgh and Philadelphia is 7.5 inches. What is the actual distance between the cities if the map scale is 1.5 inches = 60 miles?

Let *d* be the actual distance between the cities.

____ = ____   Write a proportion.

____ · $d = $ ____ · ____   Find the cross products.

$d = $ ____   Multiply.

LESSON 5-9 CONTINUED

☐ = ☐   Divide.

$d = $ ☐

The distance between the cities is ☐ miles.

Try This

1. Identify the scale factor.

	Model Aircraft	Blueprint
Length (in.)	12	2
Wing span (in.)	18	3

2. Mary's father made her a dollhouse which was modeled after the blueprint of their home. The blueprint is 24 inches by 45 inches. The scale factor is $\frac{1.5}{1}$. Find the size of the dollhouse.

3. On a road map, the distance between Dallas and Houston is 7 inches. What is the actual distance between the cities if the map scale is 1 inch = 50 kilometers?

Chapter Review

5-1 Ratios

Larry has a bag of colored dominoes. In the bag there are 20 orange dominoes, 16 red dominoes, 17 green dominoes, and 11 yellow dominoes.

Write each ratio in all three forms.

1. red dominoes to orange dominoes
2. yellow dominoes to green dominoes

3. Twelve boys showed up to football practice on Wednesday night. There are a total of 29 boys on the team. Write the ratio in all three forms of the number of boys at practice to the total number of boys on the team.

5-2 Rates

Find each rate.

4. A lawn service company charges Mrs. Smith $75 for $2\frac{1}{2}$ hours of work. What is their fee per hour?

5. A chartered bus drove 650 miles in 13 hours. What was the average rate of speed of the bus?

5-3 Slope and Rates of Change

Use the given slope and point to graph each line.

6. $\frac{1}{3}$; (2, 2)

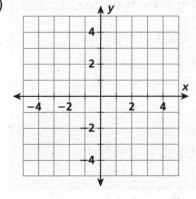

7. −2; (1, 0)

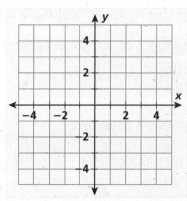

CHAPTER 5 REVIEW *CONTINUED*

5-4 Identifying and Writing Proportions

Determine whether the ratios are proportional.

8. $\frac{4}{8}, \frac{8}{10}$ 9. $\frac{2}{9}, \frac{10}{45}$ 10. $\frac{3}{5}, \frac{27}{45}$ 11. $\frac{11}{12}, \frac{22}{27}$

Complete each table of equivalent ratios.

12.

bird	8	14	22
lion		7	16

13.

pyramid	5		15	
prism	3	6		12

5-5 Solving Proportions

Use cross products to solve each proportion.

14. $\frac{4}{17} = \frac{y}{68}$ 15. $\frac{9}{x} = \frac{3}{7}$ 16. $\frac{p}{13} = \frac{33}{39}$ 17. $\frac{2}{9} = \frac{q}{54}$

18. Kathy walked 2.5 miles in 35 minutes. Use a proportion to find how long it would take her to walk 6 miles at the same speed.

5-6 Customary Measurements

Choose the most appropriate customary unit for each measurement.

19. the weight of a tractor trailer truck   20. the length of a pencil

Convert each measure.

21. $8\frac{1}{2}$ feet to inches   22. 64 quarts to gallons

23. $9\frac{1}{4}$ pounds to ounces   24. 5 miles to feet

25. A chef has 4 gallons of soup. If she serves twelve 8-ounce bowls, then how many gallons of soup are remaining?

CHAPTER 5 REVIEW CONTINUED

5-7 Similar Figures and Proportions

Use the properties of similarity to determine whether the figures are similar.

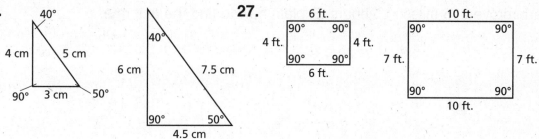

26.

27.

5-8 Using Similar Figures

28. Mrs. Nuss planted two similar rectangular gardens. Her soybean garden measures 100 yards long and 60 yards wide. Her corn garden is 150 yards long. How many yards wide is her corn garden?

29. Ned is 5 ft 4 in. tall, and casts a shadow that is 3 ft 6 in. long. At the same time his dad's tool shed casts a shadow that is 7 ft long. Estimate the height of the tool shed.

5-9 Scale Drawings and Scale Models

Identify the scale factor.

30.

	Library	Model
Height (ft)	56	4

31.

	Car	Model
Length (in.)	96	8

32. In Joe's World Atlas the distance between Cleveland and Cincinnati is $5\frac{1}{2}$ cm. What is the actual distance between the two cities if the map scale is $\frac{1}{2}$ cm = 20 miles?

CHAPTER 5 Big Ideas

Answer these questions to summarize the important concepts from Chapter 5 in your own words.

1. Darin drove 245 miles in 5 hours. Explain how to find the unit rate.

2. Explain how to graph a line using a slope of $-\frac{2}{3}$ and the point $(-4, 3)$.

3. Explain how to solve the proportion $\frac{14}{9} = \frac{x}{54}$ using cross products.

4. What is true about corresponding angles and corresponding sides of similar figures?

5. Two cities are 3.5 cm apart on a map. The scale factor is 2 cm = 25 miles. Explain how to find the actual distance d between the cities.

For more review of Chapter 5:

- Complete the Chapter 5 Study Guide and Review on pages 318–320 of your textbook.
- Complete the Ready to Go On quizzes on pages 296 and 312 of your textbook.

LESSON 6-1 Percents

Lesson Objectives
Model percents and write percents as equivalent fractions and decimals

Vocabulary
percent (p. 330)

Additional Examples

Example 1

Write the percent modeled by each grid.

A.

$\dfrac{\text{shaded}}{\text{total}} \rightarrow \dfrac{}{100} = \%$

B.

$\dfrac{\text{shaded}}{\text{total}} \rightarrow \dfrac{}{100} = \%$

LESSON 6-1 CONTINUED

Example 2

Write 28% as a fraction in simplest form.

$28\% = \dfrac{\boxed{}}{100}$ Write the percent as a fraction with a $\boxed{}$ of 100.

$= \boxed{}$ Simplify.

So 28% can be written as $\boxed{}$.

Example 3

Write 17% as a decimal.

$17\% = \boxed{}$ Write the percent as a fraction with a denominator of $\boxed{}$.

$= \boxed{}$ Divide $\boxed{}$ by 100.

Try This

1. Write the percent modeled by the grid.

2. Write 45% as a fraction in simplest form.

3. Write 67% as a decimal.

LESSON 6-2: Fractions, Decimals, and Percents

Lesson Objectives

Write decimals and fractions as percents

Additional Examples

Example 1

Write 0.7 as a percent.

0.7 = ⬚ = ⬚   Write an equivalent fraction with a denominator of ⬚.

= ⬚ %   Write the ⬚ with a percent sign.

Example 2

Write $\frac{5}{8}$ as a percent.

$\frac{5}{8} = 5 \div 8$ Use ⬚ to write the fraction as a decimal.

= ⬚

= ⬚ %   Write the decimal as a percent.

LESSON 6-2 CONTINUED

Example 3

Choose the most useful method of computation. Then solve.

If 27 out of 50 people have the newspaper delivered to their home, what percent of these people have the newspaper delivered to their home?

27 out of 50 = □/□   Think: Since the denominator is a factor of 100, _____ is a good choice.

Use _____.

$\frac{27}{50} = \frac{27 \cdot 2}{50 \cdot 2} =$ □   Write an _____ fraction with a denominator of _____

= □ %   Write the fraction as a percent.

Try This

1. Write 0.01 as a percent.

2. Write $\frac{9}{60}$ as a percent.

3. Choose the most useful method of computation. Then solve.

If 18 out of 20 dentists recommend a certain brand of toothpaste, what percent of these dentists recommend the toothpaste?

LESSON 6-3: Estimate with Percents

Lesson Objectives
Estimate percents

Additional Examples

Example 1

Use a fraction to estimate 27% of 63.

27% of 63 ≈ ☐ · 63   Think: 27% is about 25% and 25% is equivalent to ☐

≈ ☐ · 60   Change 63 to a compatible number.

≈ ☐   Multiply.

27% of 63 is about ☐.

Example 2

Tara's T's is offering 2 T-shirts for $16, while Good-T's is running their buy one for $9.99, get one for half price sale. Which store offers the better deal?

First find the discount on the second T-shirt at Good T's.

50% of $9.99 = ☐ · $9.99   Think: 50% is equivalent to ☐.

≈ ☐ · $10   Change $9.99 to a compatible number.

≈ $☐   Multiply.

The discount is approximately $☐. Since $10 + $5 = $15, the cost of two shirts at Good T's is about $☐.

The T-shirts at ☐ is a better deal.

LESSON 6-3 CONTINUED

Example 3

Use 1% or 10% to estimate the percent of each number.

A. 4% of 18

18 is about 20, so find 4% of 20.

1% of 20 = ☐

4% of 20 = 4 · ☐ = ☐   4% equals ☐ · 1%.

4% of 18 is about ☐.

B. 29% of 80

29% is about 30, so find 30% of 80.

10% of 80 = ☐

30% of 80 = 3 · ☐ = ☐   30% equals ☐ · 10%.

29% of 80 is about ☐.

Example 4

Tim spent $58 on dinner for his family. About how much money should he leave for a 15% tip?

Since $58 is about $60, find 15% of $60.

15% = 10% + 5%   Think: 15% is 10% + 5%.

10% of $60 = $ ☐

5% of $60 = $6 ÷ 2 = $ ☐   5% is $\frac{1}{2}$ of ☐ % so divide $6 by 2.

$6 + $3 = $ ☐   ☐ the 10% and 5% estimates.

Tim should leave about $ ☐ for a 15% tip.

LESSON 6-4 Percent of a Number

Lesson Objectives
Find the percent of a number

Additional Examples

Example 1

Find the percent of each number.

A. 30% of 50

　　☐ = $\frac{n}{50}$　　　　　Write a ☐.

　　☐ · 50 = ☐ · n　　　Set the cross ☐ equal. Multiply.

　　☐ = ☐ n

　　☐ = ☐　　　　　Divide each side by ☐ to isolate the variable.

　　☐ = n

30% of 50 is ☐.

B. 200% of 24

　　☐ = $\frac{n}{24}$　　　　　Write a ☐.

　　☐ · 24 = ☐ · n　　　Set the ☐ products equal. Multiply.

　　☐ = ☐ n

　　☐ = ☐　　　　　Divide each side by ☐ to isolate the variable.

　　☐ = n　　　　　200% of 24 is ☐.

LESSON 6-4 CONTINUED

Example 2

Find the percent of each number.

A. 9% of 80

9% of 80 = [ ] · 80   Write the percent as a decimal and multiply.

= [ ]

Model

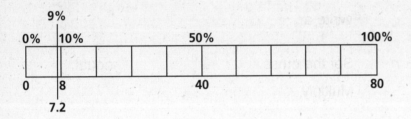

B. 3% of 12

3% of 12 = [ ] · 12   Write the percent as a decimal and multiply.

= [ ]

Example 3

The estimated world population in 2001 was 6,157 million. About 40% of the people were 19 or younger. What was the approximate number of people 19 or younger, to the nearest million?

Find 40% of 6,157 million

[ ] · 6,157   Write the percent as a decimal.

[ ]   Multiply.

The number of people 19 or under was about [ ] million.

LESSON 6-5: Solving Percent Problems

Lesson Objectives
Solve problems involving percents

Additional Examples

Example 1

Solve.

A. What percent of 40 is 25?

☐ = $\frac{25}{40}$ Write a ☐.

☐ · 40 = ☐ · 25   Set the ☐ products equal.

40 ☐ = ☐   Multiply.

☐ = ☐   Divide each side by ☐ to isolate the variable.

n = ☐

25 is ☐ % of 40.

B. 15 is 25% of what number?

☐ = $\frac{15}{n}$ Write a ☐.

n · ☐ = ☐ · 15   Set the cross ☐ equal.

☐ n = ☐   Multiply.

☐ = ☐   Divide each side by ☐ to isolate the variable.

n = ☐   15 is 25% of ☐.

LESSON 6-5 CONTINUED

Example 2

Solve.

A. 35 is 28% of what number?

$35 = 28\% \cdot n$ Write an _____.

$35 = $ _____ $\cdot n$ Write 28% as a decimal.

_____ $=$ _____   Divide each side by _____ to isolate the variable.

_____ $= n$

35 is 28% of _____.

Example 3

A portable DVD player costs $225 before tax at an appliance warehouse. What is the percent sales tax if the tax is $18?

Restate the question: What percent of 225 is 18?

_____ $=$ _____   Write a _____.

$n \cdot$ _____ $= 100 \cdot$ _____   Set the cross products equal.

$225n = $ _____   Multiply.

$n = $ _____   Divide each side by _____.

_____ % of $225 is $18. The sales tax rate is _____ %.

Try This

1. Solve.

   8 is 40% of what number?

LESSON 6-6 Percent of Change

Lesson Objectives

Solve problems involving percent of change

Vocabulary

percent of change (p. 352)

percent of increase (p. 352)

percent of decrease (p. 352)

Additional Examples

Example 1

Find each percent of change. Round answers to the nearest tenth of a percent, if necessary.

A. 65 is decreased to 38.

$65 - 38 =$ Find the amount of _____.

percent of change $= \frac{27}{65}$ Substitute values into the formula.

≈ 0.4153846 Divide.

$\approx$ ____ %   Write as a percent. Round.

The percent of decrease is about ____ %.

B. 41 is increased to 92.

$92 - 41 =$ Find the amount of change.

Percent of change $= \frac{51}{41}$ Substitute values into the formula.

≈ 1.2439 Divide.

$\approx$ ____ %   Write as a percent. Round.

The percent of increase is about ____ %.

LESSON 6-6 CONTINUED

Example 2

The regular price of a bicycle helmet is $42.99. It is on sale for 20% off. What is the sale price?

Step 1: Find the amount of the discount.

$20\% \cdot 42.99 = d$ Think: 20% of $42.99 is what number?

☐ · 42.99 = d   Write the percent as a decimal.

☐ = d

$ ☐ ≈ d   Round to the nearest cent.

The discount is $ ☐.

Step 2: Find the sale price.

regular price − amount of discount = sale price

$42.99 − $ ☐ = $ ☐

The sale price is $ ☐.

Example 3

A boutique buys hand-painted T-shirts for $12.60 each and sells them at a 110% increase in price. What is the selling price of the T-shirts?

Step 1: Find the amount n of increase.
Think: 110% of $12.60 is what number?

$110\% \cdot 12.60 = n$

☐ · 12.60 = n   Write the percent as a decimal.

☐ = n

The amount of increase is $ ☐.

LESSON 6-6 CONTINUED

Step 2: Find the selling price.
Think: retail price = wholesale price + amount of increase.

$p = \$12.60 + \$$

$p = \$$

The selling price of the hand-painted T-shirts is $ each.

Try This

1. Find the percent of change. Round answers to the nearest tenth of a percent, if necessary.

    70 is decreased to 45.

2. The regular price of a computer game is $49.88. It is on sale for 15% off.
    Find the sale price.

3. William makes T-shirts for $7.00 each and sells them after a price increase of 125%. What is the selling price of the T-shirts?

LESSON 6-7: Simple Interest

Lesson Objectives
Solve problems involving simple interest

Vocabulary
interest (p. 356)

simple interest (p. 356)

principal (p. 356)

Additional Examples

Example 1

Find each missing value.

A. $I = \_\_$, $P = \$575$, $r = 8\%$, $t = 3$ years

$I = P \cdot r \cdot t$

$I = \_\_ \cdot \_\_ \cdot \_\_$ Substitute. Use 0.08 for 8%.

$I = \$\_\_$ Multiply.

The simple interest is $\$\_\_$.

B. $I = \$204$, $P = \$1{,}700$, $r = \_\_$, $t = 6$ years

$I = P \cdot r \cdot t$

$\_\_ = \_\_ \cdot r \cdot 6$ Substitute.

$\_\_ = \_\_ \, r$ Multiply.

$\_\_ = \_\_$ Divide to isolate the variable.

$\_\_ = r$ The interest rate is __ %.

LESSON 6-7 CONTINUED

Example 2

PROBLEM SOLVING APPLICATION

Avery deposits $6,000 in an account that earns 4% simple interest. How long will it take for his account balance to reach $6,800?

1. **Understand the Problem**
   Rewrite the question as a statement:

   - Find the number of years it will take for Avery's account to reach $ _____ .

   List the important information:

   - The principal is $ _____ .

   - The interest rate is _____ %.

   - His account balance will be $ _____ .

2. **Make a Plan**
   Avery's account balance A includes the principal plus the interest:
   $A = P + I$. Once you solve for I, you can use $I = P \cdot r \cdot t$ to find the time.

3. **Solve**

   $A = P + I$

   _____ = _____ + I Substitute.

   $-6{,}000$ $-6{,}000$ Subtract to _____ the variable.

   _____ = I

   $I = P \cdot r \cdot t$

   _____ = _____ · _____ · t Substitute. Use 0.04 for 4%.

LESSON 6-7 CONTINUED

[   ] = [   ] t   Multiply.

[   ] = [   ]   Divide to isolate the

[   ] ≈ t   Round to the nearest hundredth.

It will take [   ] years.

4. Look Back

After exactly $3\frac{1}{3}$ years, Avery's money will have earned $800 in simple interest and his account balance will be $6,800.

$I = 6{,}000 \cdot 0.04 \cdot 3\frac{1}{3} = 800$

So it will take $3\frac{1}{3}$ years to reach $6,800.

Try This

1. Find the missing value.

$I = \$600$, $P = \$2{,}000$, $r =$ ■, $t = 3$ years

2. Problem Solving Application

Linda deposits $10,000 in an account that earns 8% simple interest. How long will it take for the total amount in her account to reach $12,000?

Chapter Review

6-1 Percents

Write the percent modeled by each grid.

1.
2.

Write each percent as a fraction in simplest form.

3. 27%   4. 64%   5. 35%   6. 6%

6-2 Fractions, Decimals, and Percents

Write each decimal as a percent.

7. 0.37   8. 0.045   9. 0.05   10. 0.627

11. Sam asked 20 friends if they liked peanut butter and jelly sandwiches or grilled cheese sandwiches. Thirteen of his friends said peanut butter and jelly. What percent liked peanut butter and jelly?

6-3 Estimate with Percents

Use a fraction to estimate the percent of each number.

12. 19% of 61   13. 76% of 62   14. 49% of 98   15. 19% of 86

Estimate.

16. 15% of $41.07   17. 32% of 211   18. 1% of 95

19. Alex has $15.00. He finds an item on sale for 20% off the regular price of $19.99. Does he have enough money to buy the toy? Explain.

CHAPTER 6 REVIEW CONTINUED

6-4 Percent of a Number

Find the percent of each number. If necessary, round to the nearest tenth.

20. 64% of 313　　**21.** 7% of 186　　**22.** 138% of 52

23. Hillview School has 440 students. If 55% of the students are girls, how many of the students are girls?

6-5 Solving Percent Problems

Solve. Round to the nearest tenth, if necessary.

24. 12 is what percent of 60?　　**25.** 5 is what percent of 14?

26. 9 is 75% of what number?　　**27.** 26 is 34% of what number?

6-6 Percent of Change

Find each percent of change. Round answers to the nearest tenth of a percent, if necessary.

28. 150 to 220　　**29.** 37 to 31

30. A store buys milk from a dairy for $1.90 a gallon. They sell it to their customers for $2.29 a gallon. What percent increase is this?

6-7 Simple Interest

Find each missing value.

31. $I = $ ▮, $P = \$2{,}000$, $r = 3\%$, $t = 4$ years

32. $I = \$87.50$, $P = $ ▮, $r = 5\%$, $t = 5$ years

33. $I = \$105$, $P = \$750$, $r = $ ▮, $t = 42$ months

34. Oliver deposits $500 in an account that earns 4.75% simple interest. How long will it be before the total amount is $750 dollars?

CHAPTER 6 Big Ideas

Answer these questions to summarize the important concepts from Chapter 6 in your own words.

1. Explain how to write 45% as a fraction.

2. Explain how to find 150% of 350.

3. Explain how to find the percent change when 32 is decreased to 21.

4. Explain how to find the interest rate when the simple interest is $360, the principal is $1,600, and the time is 6 years.

For more review of Chapter 6:

- Complete the Chapter 6 Study Guide and Review on pages 364–366 of your textbook.
- Complete the Ready to Go On quizzes on pages 350 and 360 of your textbook.

LESSON 7-1: Frequency Tables, Stem-Leaf Plots, and Line Plots

Lesson Objectives

Organize and interpret data in frequency tables, stem-and-leaf plots, and line plots

Vocabulary

frequency table (p. 376) _____

_____

cumulative frequency (p. 376) _____

_____

stem-and-leaf plot (p. 377) _____

_____

line plot (p. 377) _____

_____

Additional Examples

Example 1

The list shows the average high temperatures for 20 cities on one February day. Make a cumulative frequency table of the data. How many cities had average high temperatures below 59 degrees?

69, 66, 65, 51, 50, 50, 44, 41, 38, 32, 32, 28, 20, 18, 12, 8, 8, 4, 2, 2

Step 1: Look at the _____ to choose equal intervals for the data.

Step 2: Find the number of data values in each interval. Write these numbers in the "_____" column.

LESSON 7-1 CONTINUED

Step 3: Find the _____ for each row by _____ all the frequency values that are above or in that row.

February Temperatures in 20 Cities		
Average Highs	Frequency	Cumulative Frequency

_____ cities had average high temperatures below 59 degrees.

Example 2

The data shows the number of years coached by the top 15 leaders in all-time NFL coaching victories. Make a stem-and-leaf plot of the data. Then find the number of coaches who coached fewer than 25 years.

33, 40, 29, 33, 23, 22, 20, 21, 18, 23, 17, 15, 15, 12, 17

Step 1: Find the least data value and the greatest data value.

Since the data values range from _____ to _____, use tens digits for the _____ and ones digits for the _____.

Step 2: List the _____ from least to greatest on the plot.

Step 3: List the _____ for each stem from least to greatest.

LESSON 7-1 CONTINUED

Step 4: Add a _____ and a _____. The stems are the _____ digits. The leaves are the _____ digits.

_____ coaches coached fewer than 25 years.

Example 3

Make a line plot of the data. How many hours per day did Morgan babysit most often?

	M	T	W	Th	F	S	Su
Wk 1	0	6	4	6	5	8	2
Wk 2	2	7	7	7	0	6	8
Wk 3	0	6	8	5	6	1	2
Wk 4	4	8	4	3	3	6	0

Step 1. The data values range from _____ to _____. Draw a number line that includes this range.

Step 2. Put an X above the number on the number line that corresponds to the number of miles Morgan _____ each day.

The greatest number of X's appear above the number _____. This means that Morgan babysat for _____ hours most often.

Lesson 7-2: Mean, Median, Mode, and Range

Lesson Objectives
Find the mean, median, mode, and range of a data set

Vocabulary

mean (p. 381)

median (p. 381)

mode (p. 381)

range (p. 381)

outlier (p. 382)

LESSON 7-2 CONTINUED

Additional Examples

Example 1

Find the mean, median, mode, and range of the data set.

4, 7, 8, 2, 1, 2, 4, 2

mean:

4 + 7 + 8 + 2 + 1 + 2 + 4 + 2 = 30   Add the values.

30 ÷ 8 = _____   Divide the sum by the _____.

The mean is _____.

median:

1, 2, 2, 2, 4, 4, 7, 8   Arrange the values in order.

2 + 4 = 6   Since there are two middle values, find the

6 ÷ 2 = _____   _____ of these two values.

The median is _____.

mode:

1, 2, 2, 2, 4, 4, 7, 8   The value _____ occurs three times.

The mode is _____.

range:

1, 2, 2, 2, 4, 4, 7, 8   _____ the least value from the greatest value.

8 − 1 = _____

The range is _____.

LESSON 7-2 CONTINUED

Example 2

The line plot shows the number of miles each of the 20 members of the cross-country team ran in a week. Which measure of central tendency best describes these data? Justify your answer.

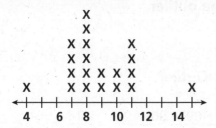

mean:

$$\frac{4 + 7 + 7 + 7 + 7 + 8 + 8 + 8 + 8 + 8 + 8 + 9 + 9 + 10 + 10 + 11 + 11 + 11 + 11 + 15}{20}$$

$= \frac{177}{20} =$ _____

median: 4, 7, 7, 7, 8, 8, 8, 8, 8, 8, 9, 9, 10, 10, 11, 11, 11, 11, 15

The median is _____.

The _____ best describes the data set because the data set has two _____ and it describes what at least half of the runners achieved.

mode:

The greatest number of X's occur above the number _____ on the line plot.

The mode is _____.

The mode focuses on one data value and does not describe the data set.

LESSON 7-2 CONTINUED

Example 3

The data shows Sara's scores for the last 5 math tests: 88, 90, 55, 94, and 89. Identify the outlier in the data set. Then determine how the outlier affects the mean, median, and mode of the data. Then tell which measure of central tendency best describes the data with the outlier.

The outlier is _____.

Without the Outlier
mean:
$$\frac{88 + 89 + 90 + 94}{4} = \frac{361}{4} = $$

median:

88, ____, 90, 94

$$\frac{89 + 90}{2} = $$

The median is _____.
mode:

There is no _____.

With the Outlier
mean:
$$\frac{55 + 88 + 89 + 90 + 94}{5} = \frac{416}{5} = $$

median:

55, 88, ____, 90, 94

The median is _____.
mode:

There is no _____.

Adding the outlier _____ the mean by _____ and the _____ by 0.5.

There is no _____ with or without the outlier. The _____ best describes the data with the outlier.

LESSON 7-3: Bar Graphs and Histograms

Lesson Objectives
Display and analyze data in bar graphs and histograms

Vocabulary

bar graph (p. 386)

double-bar graph (p. 386)

histogram (p. 387)

Additional Examples

Example 1

Use the bar graph to answer each question.

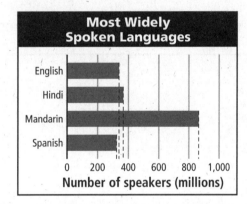

A. Which language has the least number of native speakers?

The bar for _____ is the shortest, so _____ has the least number of native speakers.

B. About how many more people speak Hindi than Spanish?

About _____ more people speak Hindi than Spanish.

LESSON 7-3 CONTINUED

Example 2

The table shows the highway speed limits on interstate roads within three states. Make a double-bar graph of the data.

State	Urban	Rural
Florida	65 mi/h	70 mi/h
Texas	70 mi/h	70 mi/h
Vermont	55 mi/h	65 mi/h

Step 1: Choose a _____ and _____ for the vertical axis.

Step 2: Draw a pair of _____ for each state's data. Use different colors to show urban and rural speed limits.

Step 3: Label the _____ and give the graph a _____.

Step 4: Make a _____ to show what each _____ represents.

LESSON 7-3 CONTINUED

Example 3

The table below shows the number of hours students watch TV in one week. Make a histogram of the data.

Number of Hours of TV			
1	//	6	///
2	////	7	//// ////
3	//// ////	8	///
4	//// /	9	////
5	//// ///		

Step 1: Make a _____ table of the data. Be sure to use equal _____.

Number of Hours of TV	Frequency

Step 2: Choose an appropriate _____ and _____ for the vertical axis. The greatest value on the scale should be at least as great as the greatest frequency.

Step 3: Draw a _____ graph for each interval. The height of the bar is the _____ for that interval. Bars must touch but not _____.

Step 4: Label the _____ and give the _____ a title.

LESSON 7-4: Reading and Interpreting Circle Graphs

Lesson Objectives
Read and interpret data presented in circle graphs

Vocabulary

circle graph (p. 390) _____

sector (p. 390) _____

Additional Examples

Example 1

Use the circle graph to answer each question.

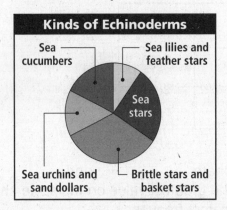

A. Which group of echinoderms includes the fewest number of species?

_____

B. Approximately what percent of echinoderm species are brittle stars and basket stars?

about _____ , so approximately _____

C. Which group is made up of a greater number of species, sea cucumbers or sea stars?

_____

LESSON 7-4 CONTINUED

Example 2

Leon surveyed 30 people about pet ownership. The circle graph shows his results. Use the graph to answer each question.

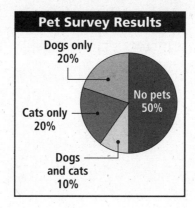

A. How many people own dogs only?

The circle graph shows that ____%, or _____, of the people

own only dogs. One-fifth of 30 is ____, so ____ people own dogs only.

B. How many people own both cats and dogs?

____ of 30 = ____ · 30

= ____

____ people own both cats and dogs.

Example 3

Decide whether a bar graph or circle graph would best display the information. Explain your answers.

A. the percent of U.S. population living in the different states

_____

B. the number of tickets sold for each night of a school play

_____

LESSON 7-5 Box-and-Whisker Plots

Lesson Objectives
Display and analyze data in box-and-whisker plots

Vocabulary

box-and-whisker plot (p. 394) _____

lower quartile (p. 394) _____

upper quartile (p. 394) _____

interquartile range (p. 395) _____

Additional Examples

Example 1

Use the data to make a box-and-whisker plot.

Heights of Basketball Players (in.)

73 67 75 81 67 75 85 69

Step 1: Order the data from least to greatest. Then find the least and

greatest _____ , the _____ , and the upper and lower

_____ .

67 67 69 73 75 75 81 85   Find the lower and upper

67 67 69 $\frac{73 + 75}{2}$ 75 81 85   Find the _____ .

= _____

67 67 69 73 | 75 75 81 85   Find the lower and upper quartiles.

LESSON 7-5 CONTINUED

first quartile = $\dfrac{67 + 69}{2}$ = ☐

third quartile = $\dfrac{75 + 81}{2}$ = ☐

Step 2: Draw a number line.

Above the number line, plot a point for each value in Step 1.

Step 3: Draw a box from the ☐ and ☐ quartiles.

Inside the box, draw a vertical line through the ☐.

Then draw the ☐ from the box to the least and greatest values.

Example 2

Use the box-and-whisker plot in Additional Example 1 and the one below to answer each question.

Heights of Baseball Players (in.)

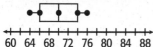

60 64 68 72 76 80 84 88

A. Which set of heights of players has the greater median?

The median of the heights of the ☐ players, about ☐ , is greater than the median of the heights of the ☐ players, about ☐ .

LESSON 7-6 Line Graphs

Lesson Objectives

Display and analyze data in line graphs

Vocabulary

line graph (p. 402)

double-line graph (p. 402)

Additional Examples

Example 1

Make a line graph of the data in the table. During which 2-hour period did the temperature change the most?

Time	11 A.M.	1 P.M.	3 P.M.	5 P.M.
Temperature	80°F	88°F	92°F	89°F

Step 1: Determine the _____ and _____ for each axis.

Place units of _____ on the horizontal axis.

Step 2: Plot a _____ for each pair of values.

Connect the points using line segments.

Step 3: Label the _____ and give the _____ a title.

The temperature changed the most between _____ and _____.

LESSON 7-6 CONTINUED

Example 2

Use the graph to estimate the population of Florida in 1950.

To estimate the population in 1950, find the point on the line between _____ and _____ that corresponds to 1950.

The graph shows about _____ people.

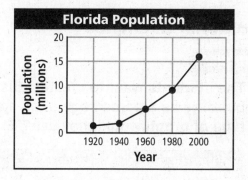

Example 3

The table shows stock prices for two stocks in one week. Make a double-line graph of the data.

	Mon	Tue	Wed	Thu	Fri
Stock A	$10	$9	$11	$10	$18
Stock B	$6	$12	$8	$14	$14

Plot a point for Stock A for each _____ of the week.

Then, using a different color, plot a point for Stock B for each _____ of the week.

_____ the points.

Make a _____ to show what each line represents.

LESSON 7-7: Choosing an Appropriate Display

Lesson Objectives

Select and use appropriate representations for displaying data

Additional Examples

Example 1

A. The students want to create a display to show each species of butterfly as a percentage of all species in the butterfly family. Which type of graph would they use? Explain.

Butterfly Family	Number of Species
Gossamer-wing	7
Skippers	10
Swallowtails	5
Whites and sulphurs	4

Each listed species is _____ of the _____ population.

A _____ shows how a set of data is divided into parts.

B. The students want to create a display to show the relationship between the species of butterflies in the park. Choose the type of graph that would best represent the data. Explain.

A _____ shows species do not _____ but are in the butterfly family.

Example 2

The table shows the number of visitors to the butterfly park during a four-month period.

Months	Visitors
May	432
June	657
July	856
August	723

178

LESSON 7-7 CONTINUED

Explain why each kind of display below would or would not appropriately represent the data.

A. Circle Graph

A circle graph shows how _____ of a set of data relates to the _____.

There is no relationship among the _____ to a _____, so a circle graph _____ appropriate.

B. Bar Graph

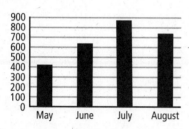

A bar graph shows _____ of data to display and show data.

The bar graph _____ the data, so a bar graph _____ appropriate.

C. Line Plot

A line plot shows the _____ of data values.

The frequencies are too _____, so a line plot _____ appropriate.

D. Line Graph

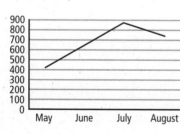

A line graph shows how data _____ over time.

The line graph _____ the data over _____, so a line graph _____ appropriate.

LESSON 7-8: Populations and Samples

Lesson Objectives
Compare and analyze sampling methods

Vocabulary

population (p. 412)

sample (p. 412)

random sample (p. 412)

convenience sample (p. 412)

biased sample (p. 413)

Additional Examples

Example 1

Determine which sampling method will better represent the entire population. Justify your answer.

Band Uniform Style

Sampling method	Results
Maria surveys only the band students she knows personally.	84% want blue uniforms
Jon writes each band student's name on a card. He questions those students whose name he draws.	61% want blue uniforms

_____ method produces results that better represent the entire band population because he uses a _____ sample.

_____ method produces results that are not as representative of the entire band population because she uses a _____ sample.

LESSON 7-8 CONTINUED

Example 2

Determine whether each sample may be biased. Explain.

A. The mayor surveys 100 supporters at a rally about the most important issues to be addressed by the city council.

The sample _____ biased. It is likely that the supporters may have _____ ideas than those not at the rally.

B. The principal sends out questionnaires to all of the students to find out what kind of music students prefer at dances.

The sample _____ biased. It is _____ because _____ student has a chance to respond.

Try This

1. Determine which sampling method will better represent the entire population. Justify your answer.

Sampling Method	Results
Pedro surveys the offense on his football team on who was the team's most valuable player	87% said the quarterback was the most valuable player.
Chad surveys 5 players from the offense and 5 players from the defense on his football team on who was the team's most valuable player	65% said the quarterback was the most valuable player.

LESSON 7-9: Scatter Plots

Lesson Objectives
Display and analyze data in scatter plots

Vocabulary

scatter plot (p. 416)

positive correlation (p. 417)

negative correlation (p. 417)

no correlation (p. 417)

Additional Examples

Example 1

Use the data to make a scatter plot. Describe the relationship between the data sets.

Step 1: Determine the scale and interval for each _____. Place the number of animals endangered in the U.S. on the

_____ axis and the number of animals

endangered in the rest of the world on the

_____ axis.

Number of Endangered Species		
Type	U.S. Only	Rest of World
Mammals	63	251
Birds	78	175
Reptiles	14	64
Amphibians	10	8
Fishes	70	11

LESSON 7-9 CONTINUED

Step 2: Plot a _____ for each pair of values.

Step 3: Label the axes and give the graph a title.

There appears to be _____ between the data sets.

Example 2

Write *positive correlation*, *negative correlation*, or *no correlation* to describe each relationship. Explain.

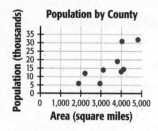

A. Population increases with area

The graph shows that as area _____, population _____. So the graph shows _____ between the data sets.

LESSON 7-10: Misleading Graphs

Lesson Objectives

Identify and analyze misleading graphs.

Additional Examples

Example 1

Which graph could be misleading? Why?

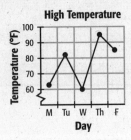

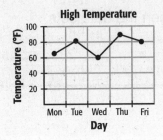

The graph on the _____ could be misleading. The _____ axis is broken, so differences in temperature appear _____.

Example 2

Explain how you could redraw the graph so it would *not* be misleading.

Draw the entire _____ scale on the graph.

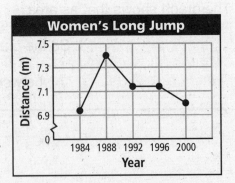

Chapter Review

7-1 Frequency Tables, Stem-and-Leaf Plots, and Line Plots

The list shows the time in minutes that students spend on the Internet.

65, 38, 44, 27, 65, 48, 52, 15, 44, 35

1. Make a cumulative frequency table of the data.

2. Make a stem-and-leaf plot of the table.

3. Make a line plot of the data.

7-2 Mean, Median, Mode, and Range

The list shows the ages of men on a tennis team.

23, 40, 42, 34, 31, 36, 49, 58, 25, 36, 28

4. Find the mean, median, and mode, and range of the data. Round your answers to the nearest tenth of a year.

5. Which measure of central tendency best represents the data? Explain.

7-3 Bar Graphs and Histograms

Use the bar graph for questions 6 and 7.

6. Which months have the most days above 80°?

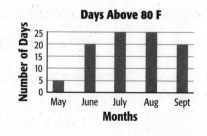

7. How many more days does it reach 80° in September compared to May?

CHAPTER 7 REVIEW CONTINUED

7-4 Reading and Interpreting Circle Graphs

Use the circle graph for problems 8-9.

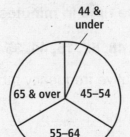

Ages of Grandparents in the U.S.

8. Are more grandparents 55-64 or 45-54?

9. Approximately what percent of grandparents are 65 & over?

7-5 Box-and-Whisker Plots

10. Make a box-and-whisker plot of the data.

    29, 25, 21, 20, 17, 16, 15, 33, 33, 30, and 15.

11. Use the box-and-whisker plot you made in Exercise 15. How many numbers are greater than the upper quartile?

7-6 Line Graphs

12. The table shows the price of two stocks over 6 months. Make a double-line graph of the data.

Month	Stock A	Stock B
January	$38.00	$50.25
February	$41.25	$49.75
March	$44.25	$48.00
April	$45.00	$45.25
May	$48.50	$46.25
June	$45.75	$48.25

7-7 Choosing an Appropriate Display

Choose the type of graph that would best represent each type of data.

13. the age of students on a junior high basketball team

14. the percent of a family's income spent on rent, food, transportation, entertainment, and savings

CHAPTER 7 REVIEW CONTINUED

7-8 Population and Samples

Determine whether each sample may be biased. Explain.

15. A school principal randomly chooses 75 students for a survey on the cafeteria food.

A reporter surveys 120 people leaving a baseball game to find out their favorite baseball team.

7-9 Scatter Plots

16. The table shows the calories and fat in selected meals at a restaurant. Use the data to make a scatter plot.

Calories	Fat
285	6
400	7
375	9
550	12
600	16
950	19
1,200	24

7-10 Misleading Graphs

17. Which graph could be misleading? Why?

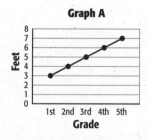
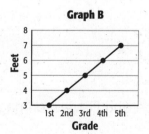

CHAPTER 7 — Big Ideas

Answer these questions to summarize the important concepts from Chapter 7 in your own words.

1. Explain how to find the mean, median, mode, and range of the data set.

   5, 8, 10, 12, 6, 5, 3

2. Explain when to use each type of display.

3. Give an example of a biased sample and explain why it is biased.

4. Explain the difference between positive and negative correlation.

For more review of Chapter 7:

- Complete the Chapter 7 Study Guide and Review on pages 430–432 of your textbook.
- Complete the Ready to Go On quizzes on pages 400 and 426 of your textbook.

LESSON 8-1 Points, Lines, and Planes

Lesson Objectives

Identify and describe geometric figures

Vocabulary

point (p. 442)

line (p. 442)

plane (p. 442)

ray (p. 442)

line segment (p. 442)

congruent (p. 442)

Additional Examples

Example 1

Identify the figures in the diagram.

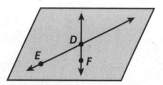

A. three points

B. two lines   Choose any   points on a line to name the line.

C. a plane   Choose any   points, not on the same line, in any order.

189   Holt Mathematics

LESSON 8-1 CONTINUED

Example 2

Identify the figures in the diagram.

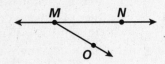

A. three rays

☐   Name the ☐ of a ray first.

B. two line segments

☐   Use the ☐ in any order to name a segment.

Example 3

Identify the line segments that are congruent in the figure.

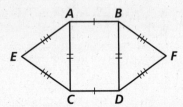

☐ ≅ ☐   One tick mark

☐ ≅ ☐   Two tick marks

☐ ≅ ☐ ≅ ☐   Three tick marks

LESSON 8-2: Classifying Angles

Lesson Objectives
Identify angles and angle pairs

Vocabulary

angle (p. 446)

vertex (p. 446)

right angle (p. 446)

acute angle (p. 446)

obtuse angle (p. 446)

straight angle (p. 446)

complementary angles (p. 446)

supplementary angles (p. 446)

LESSON 8-2 CONTINUED

Additional Examples

Example 1

Tell whether each angle is acute, right, obtuse or straight.

A.

B.

_____ angle   _____ angle

Example 2

Use the diagram to tell whether the angles are complementary, supplementary, or neither.

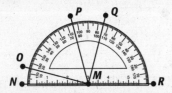

A. ∠OMP and ∠PMQ

m∠OMP = ____° and m∠PMQ = ____°

Since ____° + ____° = ____°, ∠OMP and ∠PMQ are _____.

Example 3

Angles A and B are complementary. If m∠A is 56°, what is m∠B?

m∠A + m∠B = 90°

____° + m∠B = 90°   Substitute ____° for m∠A.

−56°   −56°   Subtract ____° from both sides to isolate m∠____.

m∠B = ____°

The m∠B is ____°.

LESSON 8-3: Angle Relationships

Lesson Objectives

Identify parallel, perpendicular, and skew lines, and angles formed by a transversal

Vocabulary

perpendicular lines (p. 452)

parallel lines (p. 452)

skew lines (p. 452)

adjacent angles (p. 453)

vertical angles (p. 453)

transversal (p. 453)

corresponding angles (p. 453)

LESSON 8-3 CONTINUED

Additional Examples

Example 1

Tell whether the lines appear parallel, perpendicular, or skew.

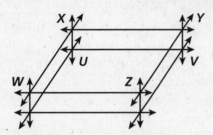

A. $\overleftrightarrow{UV}$ and $\overleftrightarrow{YV}$

$\overleftrightarrow{UV}$ ☐ $\overleftrightarrow{YV}$

The lines appear to intersect to form _____ angles.

B. $\overleftrightarrow{XU}$ and $\overleftrightarrow{WZ}$

$\overleftrightarrow{XU}$ and $\overleftrightarrow{WZ}$ are _____ .

The lines are in different planes and do not _____ .

Example 2

Line $n \parallel$ line p. Find the measure of each angle.

A. ∠2

∠2 and the 130° angle are _____ angles. Since _____ angles are congruent, m∠2 = _____ °.

B. ∠3

∠3 and the 50° angle are _____ angles. Since _____ angles are congruent, m∠3 = _____ °.

LESSON 8-4: Properties of Circles

Lesson Objectives
Identify parts of a circle and find central angle measures

Vocabulary

circle (p. 460)

center of a circle (p. 460)

radius (p. 460)

diameter (p. 460)

chord (p. 460)

arc (p. 460)

central angle (p. 461)

sector (p. 461)

Additional Examples

Example 1

Name the parts of circle M.

A. radii

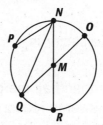

B. diameters

C. chords

LESSON 8-4 CONTINUED

Example 2

PROBLEM SOLVING APPLICATION

The circle graph shows the results of a survey about favorite types of muffins. Find the central angle measure of the sector that shows the percent of people whose favorite type of muffin is blueberry.

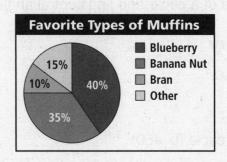

1. **Understand the Problem**

   The answer is the measure of the central angle that represents blueberry. List the important information:

   - The percent of people whose favorite muffin is blueberry is _____ %.

   - The central angle measure of the sector that represents this group is _____ % of the _____° of the circle.

2. **Make a Plan**

   There are _____° in a circle. Since the sector is 40% of the circle graph, the central angle is _____ % of the 360° in the circle.

   _____ % of 360° = 0.40 · 360°

3. **Solve**

   0.40 · 360° = _____° Multiply.

   The central angle of the sector is _____°.

4. **Look Back**

   The 40% sector is less than half the graph, and 144° is less than half of 360°. Therefore, the answer is reasonable.

LESSON 8-5 Classifying Polygons

Lesson Objectives
Identify and name polygons

Vocabulary
polygon (p. 466)

regular polygon (p. 467)

Additional Examples

Example 1

Determine whether each figure is a polygon. If it is not, explain why not.

A.

The figure _____ a polygon. It is a _____ figure with 4 line _____.

B.

The figure _____ a polygon. It is not a _____ figure.

C.

The figure _____ a polygon. The sides are _____ line segments.

D.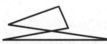

The figure _____ a polygon. The sides of a polygon cannot _____ except at endpoints.

LESSON 8-5 CONTINUED

Example 2

Name each polygon.

A.    ____ sides, ____ angles

B.    ____ sides, ____ angles

Example 3

Name each polygon and tell whether it is a regular polygon. If it is not, explain why not.

A.

The figure has ____ angles and ____ sides. It ____ a regular ____. A regular quadrilateral is also called a ____.

B.

The figure is a ____. It ____ a regular polygon because all of the sides are not ____.

Try This

1. Determine whether the figure is a polygon. If it is not, explain why not.

LESSON 8-6 Classifying Triangles

Lesson Objectives

Classify triangles by their side lengths and angle measures

Vocabulary

scalene triangle (p. 470)

isosceles triangle (p. 470)

equilateral triangle (p. 470)

acute triangle (p. 470)

obtuse triangle (p. 470)

right triangle (p. 470)

Additional Examples

Example 1

Classify each triangle according to its sides and angles.

A.

Two congruent sides

Three acute angles

This is an _____ triangle.

LESSON 8-6 CONTINUED

Classify each triangle according to its sides and angles.

B.

No congruent sides

One right angle

This is a _____ _____ triangle.

Example 2

Identify the different types of triangles in the figure, and determine how many of each there are.

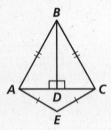

Type	How many	Name
Right		
Isosceles		
Acute		
Scalene		
Obtuse		

Try This

1. Classify the triangle according to its sides and angles.

LESSON 8-7 Classifying Quadrilaterals

Lesson Objectives

Name, identify, and draw types of quadrilaterals

Vocabulary

parallelogram (p. 474)

rectangle (p. 474)

rhombus (p. 474)

square (p. 474)

trapezoid (p. 474)

Additional Examples

Example 1

Give all the names that apply to the quadrilateral. Then give the name that best describes it.

A.

This figure has two pairs of parallel sides, so it is a

_____.

It has four congruent sides, so it is also a _____.

_____ best describes the quadrilateral.

LESSON 8-7 CONTINUED

B.

The figure has exactly one pair of opposite sides that are parallel, so it is a _____.

_____ best describes the quadrilateral.

Example 2

Draw each figure. If it is not possible to draw, explain why.

A. a rectangle that is not a square.

The figure has _____ right angles and two pairs of congruent sides. All sides _____ congruent.

B. a square that is not a rhombus

Drawing the figure is _____. All squares are _____ because they have _____ congruent sides.

LESSON 8-8 Angles in Polygons

Lesson Objectives

Find the measures of angles in polygons

Additional Examples

Example 1

Find the unknown measure in the triangle.

$80° + 55° + x =$ _____   The sum of the measures of the angles is _____°.

$135° + x =$ _____   Combine like terms.

$-$_____ $-$_____   _____ 135° from both sides.

$x =$ _____

The measure of the unknown angle is _____°.

LESSON 8-8 CONTINUED

Example 2

Find the unknown angle measure in the quadrilateral.

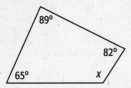

65° + 89° + 82° + x = ☐°   The sum of the measures of the angles is ☐°.

236° + x = ☐°   Combine like terms.

− 236°   − 236°   236° from both sides.

x = ☐°

The measure of the unknown angle is ☐°.

Example 3

Divide the polygon into triangles to find the sum of its angle measures.

A.

☐ · 180° = ☐°   There are ☐ triangles.

The sum of the angle measures of an octagon is ☐°.

Congruent Figures

Lesson Objectives

Identify congruent figures and use congruence to solve problems

Vocabulary

Side-Side-Side Rule (p. 484)

Additional Examples

Example 1

Identify any congruent figures.

The sides of the octagons _____ congruent. Each side of the outer figure is larger than each side of the inner figure.

Example 2

Determine whether the triangles are congruent.

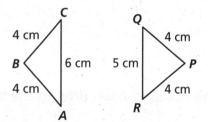

AB = ___ cm   PQ = ___ cm

BC = ___ cm   PR = ___ cm

AC = ___ cm   RQ = ___ cm

The triangles _____ congruent. Although two sides in one triangle _____ congruent to two sides in the other, the third sides _____ congruent.

LESSON 8-9 CONTINUED

Example 3

Determine the missing measures in the congruent polygons.

A.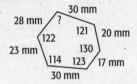

The missing angle measure is ____°.

The corresponding angles are _____.

B.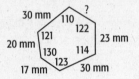

The missing side measure is ____ mm.

The corresponding sides are _____.

Try This

1. Determine whether the triangles are congruent.

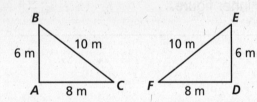

2. Determine the missing measures in the set of congruent polygons.

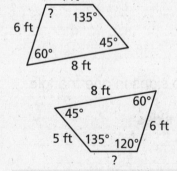

LESSON 8-10: Translations, Reflections, and Rotations

Lesson Objectives

Recognize, describe, and show transformations

Vocabulary

transformation (p. 488)

image (p. 488)

translation (p. 488)

reflection (p. 488)

line of reflection (p. 488)

rotation (p. 488)

Additional Examples

Example 1

Identify each type of transformation.

A.
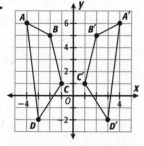

B.
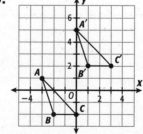

LESSON 8-10 CONTINUED

Example 2

Graph the translation of quadrilateral *ABCD* 4 units left and 2 down.

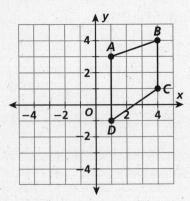

Each vertex is moved 4 units _____

and 2 units _____.

Example 3

Graph the reflection of the figure across the *x*-axis. Write the coordinates of the vertices of the image.

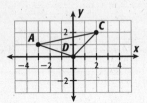

The *x*-coordinates of the corresponding

vertices are _____, and the

y-coordinates of the corresponding vertices

are _____.

The coordinates of the image are _____.

LESSON 8-10 CONTINUED

Example 4

Triangle *ABC* has vertices *A*(1, 0), *B*(3, 3), *C*(5, 0). Rotate △*ABC* 180° about the vertex *A*.

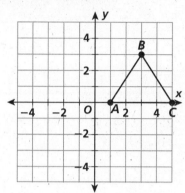

The _____ sides, $\overline{AB}$ and $\overline{A'B'}$, lie on a straight line.

Notice that the vertex *A* is the _____ of the segments $\overline{BB'}$ and $\overline{CC'}$.

Try This

1. Identify the type of transformation.

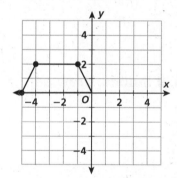

2. Graph the translation of quadrilateral *ABCD* 5 units left and 3 units down.

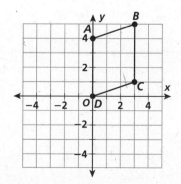

LESSON 8-11: Symmetry

Lesson Objectives
Identify symmetry in figures

Vocabulary

line symmetry (p. 494) _____

line of symmetry (p. 494) _____

asymmetry (p. 494) _____

rotational symmetry (p. 494) _____

center of rotation (p. 494) _____

Additional Examples

Example 1

Decide whether each of the figures has line symmetry. If it does, draw all the lines of symmetry.

A. _____ of symmetry

B. _____ of symmetry

LESSON 8-11 CONTINUED

Example 2

Find all the lines of symmetry in each figure.

A.   There are _____ of symmetry.

B.  There are _____ of symmetry.

Example 3

Tell how many times the figure will show rotational symmetry within one full rotation.

A.   Draw lines from the center of the figure out through identical places in the figure.

Count the number of lines drawn.

The figure will show rotational symmetry _____ times within a 360° rotation.

B.   Draw lines from the center of the figure out through identical places in the figure.

Count the number of lines drawn.

The figure will show rotational symmetry _____ times within a 360° rotation.

Chapter Review

8-1 Building Blocks of Geometry

Identify the figures in the diagram.

1. three points
2. two lines
3. a plane
4. three rays

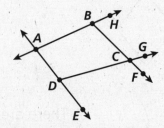

8-2 Classifying Angles

Classify each pair of angles as complementary or supplementary. Then find the missing angle measure.

5. 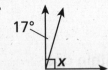   6. 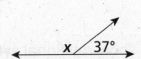   7.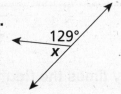

8-3 Angle Relationships

For Exercises 9-12, use the figure to complete each statement.

8. Lines *a* and *d* are ___?___.

9. Lines *b* and *a* are ___?___.

10. ∠1 and ∠5 are ___?___. They are also ___?___.

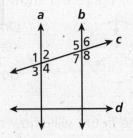

8-4 Properties of Circles

11. The circle graph shows the population of the United States in 2005, according to the U.S. Census Bureau. Find the central angle measure of the sector that shows the percent of the population that was between the ages of 0-19 that year.

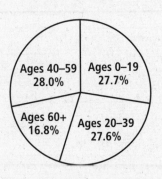

12. Name all the chords in the circle at the right.

CHAPTER 8 REVIEW CONTINUED

8-5 Classifying Polygons

Name each polygon, and tell whether it is a regular polygon. If it is not, explain why not.

12.

13.

14.

8-6 Classifying Triangles

15. Identify the different types of triangles in the figure, and determine how many of each there are.

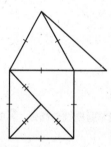

16. The sum of the lengths of the sides of triangle *ABC* is 38 in. The lengths of sides $\overline{AB}$ and $\overline{BC}$ are 14 inches and 11 inches. Find the length of side $\overline{AC}$ and classify the triangle.

8-7 Classifying Quadrilaterals

Tell whether each statement is true or false. Expain your answer.

17. All rhombuses are parallelograms.

18. Some trapezoids are parallelograms.

19. Some squares are rectangles.

CHAPTER 8 REVIEW CONTINUED

8-8 Angles in Polygons

Find the measure of the third angle in each triangle, given two angle measures. Then classify the triangle.

20. 36°, 19°

21. 61°, 52°

8-9 Congruent Figures

Determine the missing measures in each set of congruent polygons.

22.

23.

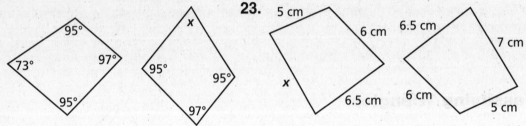

8-10 Translations, Rotations and Reflections

Identify each type of transformation.

24.

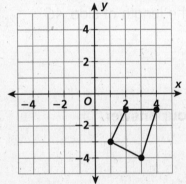

25.

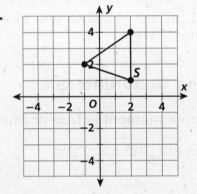

8-11 Symmetry

Find all the lines of symmetry in each flag.

26.

27.

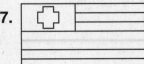

CHAPTER 8 Big Ideas

Answer these questions to summarize the important concepts from Chapter 8 in your own words.

1. Explain the difference between complementary and supplementary angles.

2. Explain the difference between a polygon and a regular polygon.

3. Two angle measures in a triangle are 35° and 77°. Explain how to find the third angle measure in the triangle.

4. Explain how to determine if two polygons are congruent.

5. Explain the difference between line symmetry and rotational symmetry.

For more review of Chapter 8:
- Complete the Chapter 8 Study Guide and Review on pages 506–508.
- Complete the Ready to Go On quizzes on pages 458, 482, and 500.

LESSON 9-1 Accuracy and Precision

Lesson Objectives
Compare the precision of measurements and determine acceptable levels of accuracy

Vocabulary

precision (p. 518) _____

accuracy (p. 518) _____

significant digits (p. 420) _____

Additional Examples

Example 1

Choose the more precise measurement in each pair.

A. 13 oz, 1 lb

Since an _____ is a smaller unit than a _____ , _____ is more precise.

B. 52 cm, 52.3 cm

Since _____ has the smaller decimal place, _____ cm is more precise.

Example 2

Determine the number of significant digits in each measurement.

A. 304.7 km

The digits 3, 4, and 7 are _____ digits, and 0 is _____ two nonzero digits.

So 304.7 has _____ significant digits.

LESSON 9-1 CONTINUED

Determine the number of significant digits in each measurement.

B. 0.0760 L

The digits 7 and 6 are _____ digits, and 0 is to the

_____ of the decimal after the last nonzero digit.

So 0.0760 L has _____ significant digits.

Example 3

Calculate 67 ft − 0.8 ft. Use the correct number of significant digits in the answer.

   67   digits to the right of the decimal point

 − 0.8   digit to the right of the decimal point

     ≈ _____ ft   Round the difference so it has no digits to the

                            _____ of the decimal point.

Try This

1. Choose the more precise measurement in the pair.

   1 gal, 5 qt

2. Determine the number of significant digits in the measurement.

   230.4 mi

3. Calculate 15 ft − 3.8 ft. Use the correct number of significant digits in the answer.

LESSON 9-2: Perimeter and Circumference

Lesson Objectives

Find the perimeter of a polygon and the circumference of a circle

Vocabulary

perimeter (p. 524)

circumference (p. 525)

Additional Examples

Example 1

Find the perimeter of the polygon.

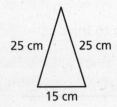

P = ☐ + ☐ + ☐   Use the side lengths.

P = ☐   Add.

The perimeter of the triangle is ☐ cm.

Example 2

Find the perimeter of the rectangle.

LESSON 9-2 CONTINUED

$P = 2l + 2w$ Use the formula.

$P = (2 \cdot) + (2 \cdot)$ Substitute for *l* and *w*.

$P = + $ Multiply.

$P = $ Add.

The perimeter of the rectangle is ____ ft.

Example 3

Find the circumference of each circle to the nearest tenth. Use 3.14 for π.

A.

$C = \pi d$ You know the diameter.

$C \approx \cdot 12$ Substitute for π and *d*.

$C \approx $ Multiply.

The circumference of the circle is about ____ in.

Example 4

The diameter of a circular pond is 42 m. What is its circumference? Use $\frac{22}{7}$ for π.

$C = \pi d$ You know the diameter.

$C \approx \cdot 42$ Substitute ____ for π and 42 for *d*.

$C \approx \cdot \frac{42}{1}$ Write 42 as a fraction.

$C \approx \cdot \frac{42}{1}$ Simplify.

$C \approx $ Multiply.

The circumference of the pond is about ____ m.

LESSON 9-3: Area of Parallelograms

Lesson Objectives

Find the area of rectangles and other parallelograms

Vocabulary

area (p. 530)

Additional Examples

Example 1

Find the area of the rectangle.

$A = lw$ Use the formula.

$A = \square \cdot \square$ Substitute for *l* and *w*.

$A = \square$ Multiply.

The area of the rectangle is _____ in².

(rectangle: 7.4 in. by 4.5 in.)

Example 2

The area of a playing field is 1470 ft and the length is 42 ft. What is the width of the field?

$A = lw$ Use the formula for the area of a rectangle.

_____ = _____ w Substitute _____ for *A* and _____ for *l*.

_____ = _____ w Divide both sides by _____ to isolate *w*.

_____ = w

The length of the garden is _____ feet.

LESSON 9-3 CONTINUED

Example 3

Find the area of the parallelogram.

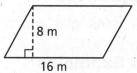

$A = bh$ Use the formula.

$A = \boxed{} \cdot \boxed{}$ Substitute for b and h.

$A = \boxed{}$

The area of the parallelogram is $\boxed{}$ m².

Example 4

A carpenter is covering a 150 ft² floor with square tiles that are each 2 ft in length. What is the least number of tiles the carpenter will need to cover the floor?

First find the area of each tile.

$A = lw$ Use the formula for the area of a square.

$A = \boxed{} \cdot \boxed{}$ Substitute 2 for l and 2 for w.

$A = \boxed{}$ Multiply.

The area of each square is $\boxed{}$ ft².

To find the number of tiles needed, divide the area of the $\boxed{}$ by the area of one $\boxed{}$.

$$\frac{\boxed{} \text{ ft}^2}{\boxed{} \text{ ft}^2} = \boxed{}$$

Since covering the floor requires more than 37 tiles, the carpet would need at least $\boxed{}$ tiles.

LESSON 9-4: Area of Triangles and Trapezoids

Lesson Objectives
Find the area of triangles and trapezoids

Additional Examples

Example 1

Find the area of each triangle.

A.

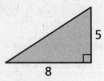

$A = \frac{1}{2}bh$ Use the formula.

$A = \frac{1}{2}(\;\square \cdot \square\;)$ Substitute ____ for b and ____ for h.

$A = \square$

The area of the triangle is ____ square units.

Example 2

Find the area of each trapezoid.

A.

$A = \frac{1}{2}h(b_1 + b_2)$ Use the formula.

$A = \frac{1}{2} \cdot \square\,(\square + \square)$ Substitute.

$A = \frac{1}{2} \cdot \square\,(\square)$ Add.

$A = \square$ Multiply.

The area of the trapezoid is ____ in².

LESSON 9-4 CONTINUED

Example 3

The state of Wisconsin is shaped somewhat like a trapezoid. What is the approximate area of the state?

$A = \tfrac{1}{2} h(b_1 + b_2)$ Use the formula.

$A = \underline{} \cdot \underline{} (\underline{} + \underline{})$ Substitute.

$A = \underline{} \cdot \underline{} (\underline{})$ Add.

$A = \underline{}$ Multiply.

The area of Wisconsin is about _____ square miles.

Try This

1. Find the area of the triangle.

2. Find the area of the trapezoid.

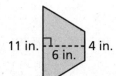

LESSON 9-5: Area of Circles

Lesson Objectives
Find the area of circles

Additional Examples

Example 1

Find the area of each circle to the nearest tenth. Use 3.14 for π.

A.

7 cm

$A = \pi r^2$ Use the formula.

$A \approx 3.14 \cdot \boxed{}^2$ Substitute $\boxed{}$ for r.

$A \approx 3.14 \cdot \boxed{}$ Evaluate the power.

$A \approx \boxed{}$ Multiply.

The area of the circle is about $\boxed{}$ cm².

Example 2

Park employees are fitting a top over a circular drain in the park. If the radius of the drain is 14 inches, what is the area of the top that will cover the drain? Use $\frac{22}{7}$ for π.

$A = \pi r^2$ Use the formula for the area of a circle.

$A \approx \frac{22}{7} \cdot \boxed{}^2$ Substitute. Use $\boxed{}$ for r.

$A \approx \frac{22}{7} \cdot \boxed{}$ Evaluate the power.

$A \approx 22 \cdot 28$

$A \approx \boxed{}$ Multiply.

The area of the top that will cover the drain is about $\boxed{}$ in².

LESSON 9-5 CONTINUED

Example 3

Use a centimeter ruler to measure the radius of the circle. Then find the area of the shaded region of the circle. Use 3.14 for π. Round your answer to the nearest tenth.

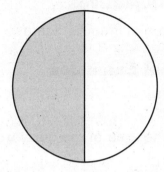

First measure the radius of the circle: It measures ____ cm.

Now find the area of the entire circle.

$A = \pi r^2$ Use the formula for the area of a circle.

$A \approx 3.14 \cdot \boxed{}^2$ Substitute. Use ____ for r.

$A \approx 3.14 \cdot \boxed{}$ Evaluate the power.

$A \approx \boxed{}$ Multiply.

$A \approx \boxed{}$

Since ____ of the circle is shaded, divide the area of the circle by ____.

____ ÷ ____ =

The area of the shaded region of the circle is about ____ cm².

Try This

1. Find the area of the circle to the nearest tenth. Use 3.14 for π.

12 ft

LESSON 9-6: Area of Irregular Figures

Lesson Objectives
Find the area of irregular figures

Additional Examples

Example 1

Estimate the area of the garden. Each square represents one square yard.

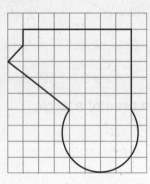

Count the number of filled or almost filled squares: ▭

Count the number of squares half-filled: ▭

Add the number of filled squares plus $\frac{1}{2}$ the number of half-filled squares:

▭ + ($\frac{1}{2} \cdot$ ▭) = ▭ + ▭ = ▭ .

The area of the garden is about ▭ yds^2.

Example 2

Find the area of the irregular figure. Use 3.14 for π.

Step 1: Separate the figure into smaller, similar figures.

Step 2: Find the area of each smaller figure.
Area of the parallelogram:
$A = bh$ Use the formula for the area of a parallelogram.

$A = \boxed{} \cdot \boxed{}$ Substitute ____ for b and ____ for h.

$A = \boxed{}$ Multiply.

Area of semicircle:
$A = \frac{1}{2}(\pi r^2)$ The area of a semicircle is $\frac{1}{2}$ the area of a circle.

$A \approx \frac{1}{2}(3.14 \cdot \boxed{}^2)$ Substitute 3.14 for π and ____ for r.

$A \approx \frac{1}{2}(\boxed{}) \approx$ Multiply.

Step 3: Add the areas to find the total area.

$A \approx \boxed{} + \boxed{} = \boxed{}$

The area of the irregular figure is about $\boxed{}$ m².

LESSON 9-6 CONTINUED

Example 3

The Wrights want to tile their entry with square foot tiles. How much tile will they need?

1. **Understand the Problem**

   Rewrite the question as a statement.

   - Find the amount of tile needed to cover the entry.

   List the important information:

   - The entry is an irregular figure.

   - The amount of tile needed is equal to the _____ of the entry.

2. **Make a Plan**

   Find the area of the entry by separating the figure into familiar figures:

   a _____ and a _____. Then add the areas of the rectangle and trapezoid to find the total area.

3. **Solve**

   Find the area of each smaller figure.

   Area of the rectangle: Area of the trapezoid:

   $A = lw$ $\qquad\qquad A = \frac{1}{2}h(b^1 + b^2)$

   $A = \boxed{} \cdot \boxed{} \qquad A = \frac{1}{2} \cdot \boxed{}(\boxed{} + \boxed{})$

   $A = \boxed{} \qquad\qquad A = \frac{1}{2} \cdot 4(\boxed{}) = \boxed{}$

   Add the areas to find the total area.

   $A = \boxed{} + \boxed{} = \boxed{}$

   The Wrights need _____ ft² of tile.

4. **Look Back**

   The area of the entry must be greater than the area of the rectangle (40 ft²), so the answer is reasonable.

LESSON 9-7: Squares and Square Roots

Lesson Objectives

Find and estimate square roots of numbers

Vocabulary

perfect square (p. 550)

square root (p. 550)

radical sign (p. 550)

Additional Examples

Example 1

Find each square.

A. Use a model. 12^2

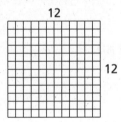

$A = lw$

$A = \boxed{} \cdot \boxed{}$ Substitute.

$A = \boxed{}$ Multiply.

The square of 12 is _____.

B. Use a calculator. 3.5^2

Press ____   .

____ = ____

The square of 3.5 is _____.

LESSON 9-7 CONTINUED

Example 2

Find each square root.

A. Method 1: Use a model. $\sqrt{16}$

The square root of 16 is ____.

B. Method 2: Use a calculator. $\sqrt{676}$

Press [2nd] [x^2] [____] [ENTER]

The square root of 676 is ____.

Example 3

Estimate each square root to the nearest whole number. Use a calculator to check your answer.

A. $\sqrt{40}$

36   40   49   Find the perfect squares nearest 40.

$\sqrt{36}$ $\sqrt{40}$ $\sqrt{49}$

 $\sqrt{40}$ Find the square roots of 36 and 49.

$\sqrt{40} \approx$ ____   40 is nearer in value to 36 than to 49.

Check

$\sqrt{40} \approx$ _____

Use a calculator to approximate $\sqrt{40}$.

6 is a reasonable estimate.

LESSON 9-8: The Pythagorean Theorem

Lesson Objectives
Use the Pythagorean Theorem to find the length of a side of a right triangle

Vocabulary
legs (p. 556)

hypotenuse (p. 556)

Pythagorean Theorem (p. 556)

Additional Examples

Example 1

Use the Pythagorean Theorem to find each missing measure.

A.

12 cm, c, 16 cm

$a^2 + b^2 = c^2$ Use the _____ Theorem.

☐² + ☐² = c^2 Substitute for a and b.

☐ + ☐ = c^2 Evaluate the powers.

☐ = c^2 Add.

☐ = $\sqrt{c^2}$ Take the square _____ of both sides.

☐ = c

The length of the hypotenuse is ☐ cm.

LESSON 9-8 CONTINUED

Example 2

PROBLEM SOLVING APPLICATION

A square field has sides of 75 feet. About how far is it from one corner of the field to the opposite corner of the field? Round your answer to the nearest tenth.

1. **Understand the Problem**
   Rewrite the question as a statement.

   • Find the _____ from one corner of the field to the opposite corner of the field.

   List the important information:

   • Drawing a segment from one corner of the field to the opposite corner of the field divides the field into two _____ triangles.

   • The segment between the two corners is the _____.

   • The sides of the field are _____, and they are each _____ feet long.

2. **Make a Plan**

   You can use the _____ Theorem to write an equation.

3. **Solve**

   $a^2 + b^2 = c^2$ Use the _____ Theorem.

   $\boxed{}^2 + \boxed{}^2 = c^2$ Substitute for the known variables.

   $\boxed{} + \boxed{} = c^2$ Evaluate the powers.

   $\boxed{} = c^2$ Add.

   $\boxed{} \approx c$ Take the square _____ of both sides.

LESSON 9-8 CONTINUED

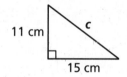 ≈ c   Round.

The distance from one corner of the field to the opposite corner is about _____ feet.

4. Look Back

The _____ is the longest side of a right triangle. Since the distance from one corner of the field to the opposite corner is greater than the length of a side of the field, the answer is reasonable.

Try This

1. Use the Pythagorean Theorem to find the missing measure.

   11 cm, 15 cm, c

2. A rectangular field has a length of 100 yards and a width of 33 yards. About how far is it from one corner of the field to the opposite corner of the field? Round your answer to the nearest tenth.

Chapter Review

9-1 Accuracy and Precision

Choose the more precise measurement in each pair.

1. 10 m, 0.01 km
2. $7\frac{1}{2}$ ft, $7\frac{3}{4}$ ft
3. 8 cm, 81 mm

9-2 Perimeter and Circumference

Find the perimeter of each polygon.

4.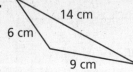

5.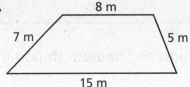

9-3 Area of Parallelograms

Find the area of each parallelogram.

6.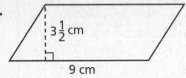

7.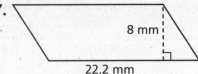

9-4 Area of Triangles and Trapezoids

Find the area of each triangle.

8.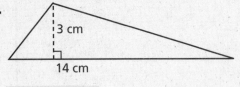

9.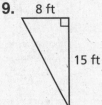

CHAPTER 9 REVIEW CONTINUED

9-5 Area of Circles

Find the area of each circle to the nearest tenth. Use 3.14 for π.

10.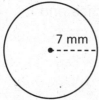

11.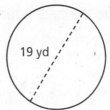

9-6 Area of Irregular Figures

Find the area of each figure. Use 3.14 for π.

12.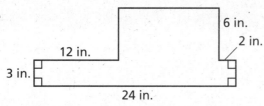

13.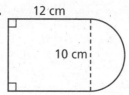

9-7 Powers and Roots

Find each square.

14. 10^2

15. 22^2

16. 13^2

Find each square root.

17. $\sqrt{121}$

18. $\sqrt{289}$

19. $\sqrt{361}$

9-8 The Pythagorean Theorem

Use the Pythagorean Theorem to find each missing length.

20.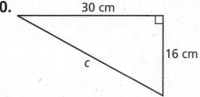

21.

CHAPTER 9 Big Ideas

Answer these questions to summarize the important concepts from Chapter 9 in your own words.

1. Explain how to find the perimeter of a rectangle with length 19 inches and width 14 inches.

2. Explain how to find the area of a circle with diameter 16 yards.

3. Explain how to estimate $\sqrt{46}$.

4. The lengths of the legs of a right triangle are 3 meters and 4 meters. Explain how to find the length of the hypotenuse using the Pythagorean Theorem.

For more review of Chapter 9:

- Complete the Chapter 9 Study Guide and Review on pages 566–568 of your textbook.
- Complete the Ready to Go On quizzes on pages 546 and 560 of your textbook.

LESSON 10-1: Introduction to Three-Dimensional Figures

Lesson Objectives

Identify various three-dimensional figures

Vocabulary

face (p. 580)

edge (p. 580)

polyhedron (p. 580)

vertex (p. 580)

base (p. 580)

prism (p. 580)

pyramid (p. 580)

cylinder (p. 581)

cone (p. 581)

LESSON 10-1 CONTINUED

Additional Examples

Example 1

Identify the base or bases of the solid. Then name the solid.

A.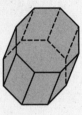

There are two bases, and they are both _____.

The other faces are _____.

The figure is an _____.

Example 2

Classify each figure as a polyhedron or not a polyhedron. Then name the figure.

A.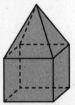

The figure _____ a polyhedron.

The figure is made up of a rectangular _____ and a rectangular _____.

LESSON 10-2: Volume of Prisms and Cylinders

Lesson Objectives
Find the volume of prisms and cylinders

Vocabulary
volume (p. 586) _____

Additional Examples

Example 1

Find how many cubes the prism holds. Then give the prism's volume.

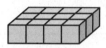

You can find the volume of this prism by counting how many cubes tall, long, and wide the prism is and then multiplying.

☐ · ☐ · ☐ = ☐

There are ☐ cubes in the prism, so the volume is ☐ cubic units.

Example 2

Find the volume of the prism to the nearest tenth.

4 ft, 4 ft, 12 ft

$V = Bh$ Use the formula.

The bases are _____.

The area of each rectangular

base is ☐ · ☐ = ☐.

$V = 48 \cdot$ ☐   Substitute for B and h.

$V =$ ☐   Multiply.

The volume to the nearest tenth is ☐ ft³.

LESSON 10-2 CONTINUED

Example 3

Find the volume of a cylinder to the nearest tenth. Use 3.14 for π.

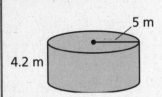

$V = \pi r^2 h$ Use the formula.

The radius of the cylinder is ____ m, and the height is ____ m.

$V \approx 3.14 \cdot \boxed{}^2 \cdot 4.2$ Substitute for r and h.

$V \approx \boxed{}$ Multiply.

The volume is about ____ m³.

Try This

1. Find how many cubes the prism holds. Then give the prism's volume.

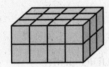

2. Find the volume of the prism to the nearest tenth.

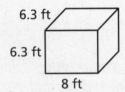

3. Find the volume of the cylinder to the nearest tenth. Use 3.14 for π.

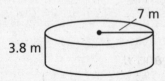

LESSON 10-3: Volume of Pyramids and Cones

Lesson Objectives

Find the volume of pyramids and cones

Additional Examples

Example 1

Find the volume of the pyramid to the nearest tenth.

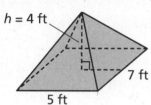

$V = \frac{1}{3}Bh$ Use the formula.

$B = \quad\cdot\quad =$ Find the area of the rectangular base.

$V = \frac{1}{3} \cdot \quad \cdot$ Substitute for B and h.

$V =$ Multiply.

The volume is _____ ft³.

Example 2

Find the volume of a cone to the nearest tenth. Use 3.14 for π.

A. 6 yd, 5 yd

$V = \frac{1}{3}\pi r^2 h$ Use the formula.

$V \approx \frac{1}{3} \cdot \quad \cdot \quad^2$ Substitute.

$V \approx$ Multiply.

The volume is about _____ yd³.

LESSON 10-4: Surface Area of Prisms and Cylinders

Lesson Objectives

Find the surface area of prisms and cylinders

Vocabulary

net (p. 597)

surface area (p. 597)

Additional Examples

Example 1

Find the surface area of the prism formed by the net.

$S = 2lw + 2lh + 2wh$

$S = (2 \cdot \boxed{} \cdot \boxed{}) + (2 \cdot \boxed{} \cdot \boxed{}) +$

$\quad\quad (2 \cdot \boxed{} \cdot \boxed{})$ Substitute.

$S = \boxed{} + \boxed{} + \boxed{}$ Multiply.

$S = \boxed{}$ Add.

The surface area of the prism is  in².

Example 2

Find the surface area of the cylinder formed by the net to the nearest tenth. Use 3.14 for π.

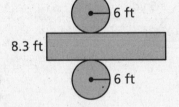

$S = 2\pi r^2 + 2\pi rh$ Use the formula.

$S \approx (2 \cdot \boxed{} \cdot \boxed{}^2) +$

$\quad\quad (2 \cdot \boxed{} \cdot \boxed{} \cdot \boxed{})$ Substitute.

LESSON 10-4 CONTINUED

S ≈ ☐ + ☐   Multiply.

S ≈ ☐   Add.

S ≈ ☐   Round.

The surface area of the cylinder is about ☐ ft².

Example 3

PROBLEM SOLVING APPLICATION

What percent of the total surface area of the soup can is covered by the label?

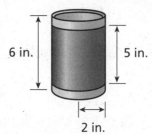

1. **Understand the Problem**

   Rewrite the question as a statement:

   • Find the _____ of the label, and compare to the total _____ of the can.

   List the important information:

   • The can is _____ shaped.

   • The height of the can is _____ .

   • The radius of the can is _____ .

   • The height of the label is _____ .

2. **Make a Plan**

   Find the _____ of the can and the _____ of the label. _____ to find the percent of the surface area covered by the label.

LESSON 10-4 CONTINUED

3. Solve

$S = 2\pi r^2 + 2\pi rh$

$\approx (2 \cdot 3.14 \cdot 2^2) + (2 \cdot 3.14 \cdot 2 \cdot 6)$ Substitute for r and h.

$\approx$ _____ in²

$A = lw$
$= (2\pi r)w$ Substitute $2\pi r$ for l.

$\approx (2 \cdot$ _____ $\cdot$ _____$)($ _____ $)$ Substitute for r and w.

$\approx$ _____ in²

Percent of the surface covered by the label: ─── = ─── %.

About _____ % of the can's surface is covered by the label.

4. Look Back
Estimate and compare the areas of the two areas.

Try This

1. Find the surface area of the prism formed by the net.

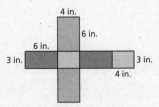

2. Find the surface area of the cylinder formed by the net to the nearest tenth. Use 3.14 for π.

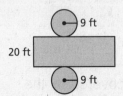

LESSON 10-5: Changing Dimensions

Lesson Objectives
Find the volume and surface area of similar three-dimensional figures

Additional Examples

Example 1

A. The surface area of a box is 35 in^2. What is the surface area of a similar box that is larger by a scale factor of 7?

$S = 35 \cdot 7^2$ Use the _____ area of the smaller box and the square of the scale _____.

$S = 35 \cdot$ _____   Evaluate the power.

$S =$ _____   Multiply.

The surface area of the larger box is _____ in^2.

Example 2

The volume of a child's pool is 28 ft^3. What is the volume of a similar pool that is larger by a scale factor of 4?

$V =$ _____ $\cdot$ _____^{3} Use the _____ of the smaller pool and the _____ of the scale factor.

$V =$ _____ $\cdot$ _____   Evaluate the power.

$V =$ _____   Multiply.

The volume of the larger pool is _____ ft^3.

LESSON 10-5 CONTINUED

Example 3

PROBLEM SOLVING APPLICATION

The sink in Kevin's workshop measures 16 in. by 15 in. by 6 in. Another sink with a similar shape is larger by a scale factor of 2. There are 231 in^3 in 1 gallon. Estimate how many more gallons the larger sink holds.

1. **Understand the Problem**

    Rewrite the question as a statement.

    - Compare the capacities of 2 similar sinks and estimate how much more water the larger sink holds.

    List the important information:

    - The small sink is 16 in. × 15 in. × 6 in.

    The larger sink is similar to the small sink by a scale factor of ____.

    - ____ in^3 = 1 gal

2. **Make a Plan**

    You can write an equation that relates the volume of the large sink to the volume of the small sink. Volume of large sink = Volume of small sink × (____)^3. Then convert cubic inches to ____.

3. **Solve**

    Volume of small sink = 16 × 15 × 6 = ____ in^3

    Volume of large sink = ____ × ____^{3} = 11,520 in^3

    Convert each volume into gallons:

    1,440 in^3 × $\frac{1 \text{ gal}}{231 \text{ in}^3}$ ≈ ____ gal   11,520 in^3 × $\frac{1 \text{ gal}}{231 \text{ in}^3}$ ≈ ____ gal

    Subtract capacities: 50 gal − 6 gal = ____ gal

    The large sink holds about ____ gallons more than the small sink.

4. **Look Back**

    Double the dimensions of the small sink and find the volume:

    32 × 30 × 12 = ____ in^3.

    Subtract the volumes of the two sinks: ____ − 1,440 = ____ in^3.

    Convert this measurement to gallons: 10,080 × $\frac{1 \text{ gal}}{231 \text{ in}^3}$ ≈ ____ gal.

Chapter Review

10-1 Introduction to Three-Dimensional Figures

Classify each figure as a polyhedron or not a polyhedron. Then name the figure.

1.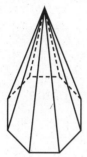
2.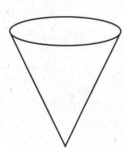
3.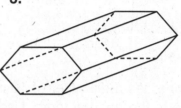

Classify each figure as a polyhedron or not a polyhedron. Then name the figure.

4. one rectangular base and four triangular faces

5. two parallel, congruent triangular bases and three other polygonal faces

10-2 Volume of Prisms and Cylinders

Find the volume of each figure to the nearest tenth. Use 3.14 for π.

6.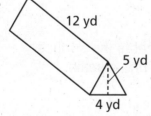
7.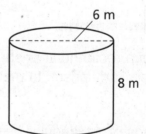

8. A shipping box is shaped like a rectangular prism. It is 12 in. long, 7 in. wide, and 4 in. high. Find its volume.

9. A can of bubbles is shaped like a cylinder. It is 5 cm wide and 13 cm tall. Find its volume to the nearest tenth. Use 3.14 for π.

CHAPTER 10 REVIEW CONTINUED

10-3 Volume of Pyramids and Cones

Find the volume of each pyramid to the nearest tenth.

10.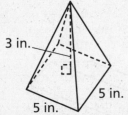

11.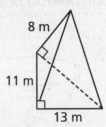

12. A cone has a diameter of 10 in. and a height of 7 in. Find the volume of the cone to the nearest tenth. Use 3.14 for π.

10-4 Surface area of Prisms and Cylinders

Find the surface area of the prism formed by the net.

13.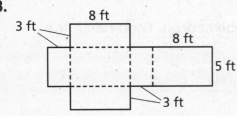

14.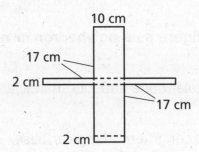

15. A can of play dough is cylindrical. The can is 7 cm wide and 7 cm tall. If the label on the can is 3 cm tall and goes all around the can, how much paper is needed for the label to the nearest tenth? Use 3.14 for π.

10-5 Changing Dimensions

16. The surface area of a rectangular prism is 14,500 cm². What is the surface area of a smaller, similarly prism, to the nearest tenth that has a scale factor of $\frac{1}{6}$?

17. A tent is in the shape of a triangular prism. The volume of the tent is 176 ft³. What is the volume of a larger, similarly shaped tent that has a scale factor of 3?

18. The surface area of a cylinder is 678 in.². Its volume is about 1,356 in.³. What are the surface area and volume of a similarly shaped cylinder that has a scale factor of 5?

CHAPTER 10 Big Ideas

Answer these questions to summarize the important concepts from Chapter 10 in your own words.

1. Explain why cylinders and cones are not polyhedrons.

2. Explain how to find the volume of a cylinder with radius 8 inches and height 6 inches.

3. Explain how to find the surface area of a rectangular prism with length 3 feet, width 6 feet, and height 7 feet.

4. The surface area of a box is 34 cm². Explain how to find the surface area of a similar box that is larger by a scale factor of 9.

For more review of Chapter 10:

- Complete the Chapter 10 Study Guide and Review on pages 616–618 of your textbook.
- Complete the Ready to Go On quizzes on pages 594 and 610 of your textbook.

LESSON 11-1 Probability

Lesson Objectives
Use informal measures of probability

Vocabulary

experiment (p. 628)

trial (p. 628)

outcome (p. 628)

event (p. 628)

probability (p. 628)

complement (p. 629)

Additional Examples

Example 1

Determine whether each event is impossible, unlikely, as likely as not, likely, or certain.

A. rolling an odd number on a number cube
   There are 6 possible outcomes:

Odd	*Not* Odd
1, 3, 5	2, 4, 6

_____ of the outcomes are odd.

Rolling an odd number is _____.

LESSON 11-1 CONTINUED

B. rolling a number less than 2 on a number cube
There are 6 possible outcomes:

Less than 2	*Not* Less than 2
1	2, 3, 4, 5, 6

Only ☐ of the outcomes is less than 2.

Rolling a number less than 2 is ☐.

Example 2

A bag contains circular chips that are the same size and weight. There are 8 purple, 4 pink, 8 white, and 2 blue chips in the bag. The probability of drawing a pink chip is $\frac{2}{11}$. What is the probability of not drawing a pink chip?

$P(\text{event}) + P(\text{complement}) =$ ☐

$P(\text{pink}) + P(\text{not pink}) =$ ☐

☐ $+ P(\text{not pink}) =$ ☐   Substitute ☐ for $P(\text{pink})$.

$-$ ☐ $\qquad -$ ☐   Subtract ☐ from both sides.

$P(\text{not pink}) =$ ☐

The probability of not drawing a pink marble is ☐.

Example 3

Mandy's science teacher almost always introduces a new chapter by conducting an experiment. Mandy's class finished a chapter on Friday. Should Mandy expect the teacher to conduct an experiment next week? Explain.

Since the class just finished a chapter, they will be starting a new chapter.

It is ☐ the teacher will conduct an experiment.

LESSON 11-2: Experimental Probability

Lesson Objectives
Find experimental probability

Vocabulary
experimental probability (p. 632)

Additional Examples

Example 1

During skating practice, Sasha landed 7 out of 12 jumps. What is the experimental probability that she will land her next jump?

$$P \approx \frac{\text{number of times an event occurs}}{\text{total number of trails}}$$

$$P(\text{land}) \approx \frac{\text{number of jumps}}{\text{number of jumps}}$$

$\approx$ _____ Substitute.

The experimental probability that Sasha will land her next jump is _____.

Example 2

Students have checked out 55 books from the library. Of these, 32 books are fiction.

A. What is the experimental probability that the next book checked out will be fiction?

$$P(\text{fiction}) \approx \frac{\text{number of _____ books checked out}}{\text{total number of books checked out}}$$

$\approx$ _____ Substitute.

The experimental probability that the next book checked out will be fiction is approximately _____.

LESSON 11-3: Make a List to Find Sample Spaces

Lesson Objectives

Use counting methods to determine possible outcomes

Vocabulary

sample space (p. 636)

Fundamental Counting Principle (p. 637)

Additional Examples

Example 1

PROBLEM SOLVING APPLICATION

One bag has a red tile, a blue tile, and a green tile. A second bag has a red tile and a blue tile. Vincent draws one tile from each bag. What are all the possible outcomes? How many outcomes are in the sample space?

1. **Understand the Problem**
   Rewrite the question as a statement.

   - Find all the possible _____ of drawing one tile from each bag, and determine the size of the _____ space.

   List the important information:

   - There are ____ bags.

   - One bag has a ____ tile, a ____ tile, and a ____ tile.

   - The other bag has a ____ tile and a ____ tile.

LESSON 11-3 CONTINUED

2. **Make a Plan**
   You can make an organized list to show all possible outcomes.

3. **Solve**
   Let R = red tile, B = blue tile, and G = green tile.
   Record each possible outcome.

   The possible outcomes are

   ____ , ____ , ____ , ____ , ____ ,

   and ____ . There are ____ possible outcomes in the sample space.

Bag 1	Bag 2

4. **Look Back**
   Each possible outcome that is recorded in the list is different.

Example 2

There are 4 cards and 2 tiles in a board game. The cards are labeled N, S, E, and W. The tiles are numbered 1 and 2. A player randomly selects one card and one tile. What are all the possible outcomes? How many outcomes are in the sample space?

You can make a ____ diagram to show the sample space.

List each letter of the cards. Then list each color of the tiles.

     N   S   E   W

There are ____ possible outcomes in the sample space.

LESSON 11-3 CONTINUED

Example 3

Carrie rolls two 1-6 number cubes. How many outcomes are possible?

The first number cube has ____ outcomes.

The second number cube has ____ outcomes.

____ • ____ = ____   Use the Fundamental Counting Principle.

There are ____ possible outcomes.

Try This

1. Darren has two bags of marbles. One has a green marble and a red marble. The second bag has a blue and a red marble. Darren draws one marble from each bag. What are all the possible outcomes? How many outcomes are in the sample space?

2. There are 2 marbles and 3 cubes in a board game. The marbles are black and red. The cubes are numbered 1, 2, and 3. A player randomly selects one marble and one cube. What are all the possible outcomes? How many outcomes are in the sample space?

1

2

3

3. Juan tosses a coin and rolls a number cube. How many outcomes are possible?

LESSON 11-4

Theoretical Probability

Lesson Objectives
Find the theoretical probability of an event

Vocabulary
theoretical probability (p. 640)

Additional Examples

Example 1

Andy has 20 marbles in a bag. Of these, 9 are clear and 11 are blue. Find the probability of each event. Write your answer as a fraction, a decimal, and a percent.

$$P = \frac{\text{number of _____ outcomes}}{\text{total number of _____ outcomes}}$$

$P(\text{clear}) = \dfrac{\text{number of _____ marbles}}{\text{total number of marbles}}$ Write the ratio.

= _____   Substitute.

= _____ = _____ %   Write as a decimal and write as a percent.

The theoretical probability of drawing a clear marble is _____ , _____ , or _____ %.

The theoretical probability of drawing a blue marble is 1 − _____ , or _____ , _____ , or _____ %.

LESSON 11-4 CONTINUED

Example 2

There are 13 boys and 10 girls on the track team. The name of each team member is written on an index card. A card is drawn at random to choose a student to run a sprint and the card is replaced in the stack.

A. Find the theoretical probability of drawing a boy's name.

$$P(\text{boy}) = \frac{\text{number of _____ on the team}}{\text{number of _____ on the team}}$$ Find the theoretical probability.

= ____ Substitute.

Try This

1. Find the probability. Write your answer as a fraction, as a decimal, and as a percent. Jane has 20 marbles in a bag. Of these 8 are green. Find the probability of drawing a green marble from the bag.

2. There are 15 boys and 12 girls in the class. Find the theoretical probability of drawing a boy's name.

LESSON 11-5: Probability of Independent and Dependent Events

Lesson Objectives
Find the probability of independent and dependent events

Vocabulary
independent events (p. 648)

dependent events (p. 648)

Additional Examples

Example 1

Decide whether each set of events are dependent or independent. Explain your answer.

A. Kathi draws a 4 from a set of cards numbered 1–10 and rolls a 2 on a number cube.

Since the outcome of drawing the card does not _____ the outcome of rolling the cube, the events are _____.

B. Yuki chooses a book from the shelf to read, and then Janette chooses a book from the books that remain.

Since Janette cannot pick the same book that Yuki picked, and since there are fewer books for Janette to choose from after Yuki chooses, the events are _____.

LESSON 11-5 CONTINUED

Example 2

Find the probability of choosing a green marble at random from a bag containing 5 green and 10 white marbles and then flipping a coin and getting tails.

The outcome of choosing the marble does not _____ the

outcome of flipping the coin, so the events are _____.

P(green and tails) = P(green) · P(tails)

   = ☐ · ☐

The probability of choosing a green marble and a coin landing on tails

is ☐.

Example 3

A reading list contains 5 historical books and 3 science-fiction books. What is the probability that Juan will randomly choose a historical book for his first report and a science-fiction book for his second?

The first choice changes the number of books left, and may change the

number of science-fiction books left, so the events are _____.

P(historical) = ☐   There are ☐ historical books out

                           of ☐ books.

P(science-fiction) = ☐   There are ☐ science-fiction books

                           left out of ☐ books.

P(historical and then science-fiction) = P(A) · P(B after A)

   = ☐ · ☐   Multiply.

The probability of Juan choosing a historical book and then choosing a

science-fiction book is ☐.

LESSON 11-6 Combinations

Lesson Objectives

Find the number of possible combinations

Vocabulary

combination (p. 652)

_____

Additional Examples

Example 1

Kristy's Diner offers customers a choice of 4 side dishes with each order: carrots, corn, french fries, and mashed potatoes. How many different combinations of 3 side dishes can Kareem choose?

Begin by listing all the _____ choices of side dishes taken three at a time.

Because _____ does not matter, you can eliminate repeated triples. For example 1, 2, 3 is already listed, so 2, 1, 3 can be eliminated.

1, 2, 3	2, 1, 3	3, 1, 2	4, 1, 2
1, 2, 4	2, 1, 4	3, 1, 4	4, 1, 3
1, 3, 4	2, 3, 4	3, 2, 4	4, 2, 3

There are ____ possible combinations of 3 side dishes Kareem can choose with his order.

LESSON 11-6 CONTINUED

Example 2

PROBLEM SOLVING APPLICATION

Lara is going to make a double-dip cone from a choice of vanilla, chocolate, and strawberry. She wants each dip to be a different flavor. How many different cone combinations can she choose?

1. **Understand the Problem**
   Rewrite the question as a statement.

   - Find the number of possible _____ of two flavors Lara can choose.

   List the important information:

   - There are _____ flavor choices in all.

2. **Make a Plan**

   You can make a _____ diagram to show the possible combinations.

3. **Solve**

   The tree diagram shows _____ possible ways to combine two flavors, but each combination is listed twice. So there are _____ ÷ _____ = _____ possible combinations.

4. **Look Back**
   You can also check by making a list. The vanilla can be paired with two other flavors and the chocolate with one. The total number of possible pairs is 2 + 1 = 3.

LESSON 11-7 Permutations

Lesson Objectives
Find the number of possible permutations

Vocabulary
permutation (p. 656)

factorial (p. 656)

Additional Examples

Example 1

In how many different orders can you arrange the letters *A*, *B*, and *T*?

Use a list to find the possible _____.

There are ____ ways to order the letters.

LESSON 11-7 CONTINUED

Example 2

Mary, Rob, Carla, and Eli are lining up for lunch. In how many different ways can they line up for lunch?

Once you fill a position, you have one less choice for the next position.

There are [4] choices for the first position.

There are [3] remaining choices for the second position.

There are [2] remaining choices for the third position.

There is [1] choice left for the fourth position.

[4] · [3] · [2] · [1] = [24]   Multiply.

There are [24] different ways the students can line up for lunch.

Example 3

How many different orders are possible for Shellie to line up 8 books on a shelf?

Number of permutations = 8!

= [8] · [7] · [6] · [5] · [4] · [3] · [2] · [1]

= [40,320]

There are [40,320] different ways for Shellie to line up 8 books on the shelf.

Chapter Review

11-1 Probability

Determine whether each event is impossible, unlikely, as likely as not, likely, or certain.

1. rolling a 7 on a number cube

2. flipping a coin and getting heads

3. drawing a red marble from a bag with 7 blue marbles and 10 red marbles

4. Charlie rolls two number cubes. The probability the sum is less than 4 is $\frac{2}{21}$. What is the probability of having a sum of 4 or greater?

5. Anna exercises for at least 45 minutes on days she has to work. If it is Monday and Anna has to work, would you expect her to exercise over 30 minutes? Explain.

11-2 Experimental Probability

6. David made 16 out of 28 free throws at basketball practice. What is the experimental probability of making his next free throw?

7. Christina scored an A on 7 out of 10 math quizzes this quarter. What is the experimental probability that she scores an A on her next math quiz?

8. For the past two weeks, Courtney has picked 9 long sleeve and 6 short sleeve shirts to wear to school.

   a) What is the probability that the next shirt she picks will be short sleeve?

   b) What is the probability that the next shirt she picks will be long sleeve?

CHAPTER 11 REVIEW CONTINUED

11-3 Make a List to Find Sample Space

9. Josh and Jennifer are playing a game with a spinner and a coin. The spinner is divided into 6 equal sections. A turn consists of one spin of the spinner and one flip of the coin. What are all the possible outcomes? How many outcomes are in the sample space?

10. Chad has blue pants and tan pants. He also has a white shirt, gray shirt, blue shirt and a yellow shirt. What are all the possible outcomes for an outfit? How many outcomes are in the sample space?

11. McKenzie has a sundae bar for her birthday party. It has vanilla, chocolate and strawberry ice cream. For toppings it has hot fudge, caramel sauce, and marshmallow topping. What are all the possible outcomes for a sundae? How many outcomes are in the sample space?

11-4 Theoretical Probability

Find the probability of each event. Write your answer as a fraction, as a decimal, and as a percent.

12. randomly drawing a heart from a shuffled deck of 52 cards with 13-card suits: diamonds, hearts, clubs and spades

13. randomly drawing one of the 4-D Scrabble tiles from a complete set of 100 Scrabble Tiles

A twelve-sided number cube is rolled. What is the probability of each event?

14. P(even #)

15. P(greater than 10)

16. P(less than 6)

17. P(9)

18. Lucy and her cousins are drawing names to buy gifts for each other. She has 6 boy cousins and 7 girl cousins. If Lucy is randomly drawing one of her cousin's names, what is the probability she draws a girl?

CHAPTER 11 REVIEW CONTINUED

11-5 Probability of Independent and Dependent Events

Decide whether each set of events is independent or dependent. Explain.

19. A man chooses a movie at the video store and then chooses a second movie from those remaining.

20. A child takes a coin out of his piggybank and then picks another one after replacing the first coin.

21. Julio has a bag of marbles that contains 6 red, 11 blue, 5 green, and 10 yellow. What is the probability Julio picks a green marble first and then picks a red marble without replacing the green marble?

11-6 Combinations

22. Gino's pizza offers 5 toppings, pepperoni, sausage, mushrooms, green peppers, and extra cheese, for their pizzas. How many different 3 topping pizzas can be made?

23. Graham, Lupe, Chandra and Karl are going to play checkers. How many different ways can they pair up?

24. A student takes 2 electives in a year. How many different combinations of 2 electives can be formed from 10 elective choices?

11-7 Permutations

25. A family of 5 is posing for a family picture. If they want to stand in a line, how many different orders can they make?

26. Students in a computer class need to create a 4-digit password using the numbers 1-9 without repeating. How many different passwords can be created?

27. Find the number of permutations of the letters in the word "MATH"?

Determine whether each problem involves combinations or permutations. Explain your answer.

28. Picking 6 people to play a game out of 10 people.

29. The seating order in your math class

CHAPTER 11 Big Ideas

Answer these questions to summarize the important concepts from Chapter 11 in your own words.

1. Tony answered 27 out of 30 questions correctly. Explain how to find the experimental probability that Tony will answer the next question correctly.

2. Explain how to find the theoretical probability of rolling a number greater than 4 on a fair number cube.

3. Explain the difference between combinations and permutations.

For more review of Chapter 11:

- Complete the Chapter 11 Study Guide and Review on pages 664–666 of your textbook.
- Complete the Ready to Go On quizzes on pages 646 and 660 of your textbook.

LESSON 12-1: Solving Two-Step Equations

Lesson Objectives
Solve two-step equations

Additional Examples

Example 1

Solve.

$9c + 3 = 39$

$-\underline{} \quad -\underline{}$ Subtract ___ from both sides.

$9c = $

$ = $ Divide both sides by ___.

$c = $

Example 2

Solve.

$6 + \dfrac{y}{5} = 21$

$-\underline{} \quad -\underline{}$ Subtract ___ from both sides.

$\dfrac{y}{5} = $

$()\dfrac{y}{5} = ()15$ Multiply both sides by ___.

$y = $

LESSON 12-1 CONTINUED

Example 3

Jamie rented a canoe while she was on vacation. She paid a flat rental fee of $85.00 plus $7.50 each day. Her total cost was $130.00. For how many days did she rent the canoe?

Let *d* represent the number of days she rented the canoe.

$7.5d + 85 = 130$

— ____ — ____   Subtract ____ from both sides.

$7.5d = $ ____

____ = ____   Divide both sides by ____ .

$d = $ ____

Jamie rented the canoe for ____ days.

Try This

1. Solve.

$-6m - 8 = -50$

2. Solve.

$8 + \dfrac{y}{2} = 48$

3. Jack's father rented a car while they were on vacation. He paid a rental fee of $20.00 per day plus 20¢ a mile. He paid $25.00 for mileage and his total bill for renting the car was $165.00. For how many days did he rent the car?

LESSON 12-2: Solving Multi-Step Equations

Lesson Objectives
Solve multi-step equations

Additional Examples

Example 1

Solve $12 - 7b + 10b = 18$.

$12 - 7b + 10b = 18$

$12 + \boxed{}b = 18$ Combine $\boxed{}$ terms.

$\underline{-\boxed{}} \quad \underline{\boxed{}}$ Subtract $\boxed{}$ from both sides.

$3b = \boxed{}$

$\dfrac{\boxed{}}{} = \boxed{}$ Divide both sides by $\boxed{}$.

$b = \boxed{}$

Example 2

Solve $5(y - 2) + 6 = 21$.

$5(y - 2) + 6 = 21$

$5() - 5() + 6 = 21$ Distribute $\boxed{}$ on the left side.

$5y - \boxed{} = 21$ Simplify.

$\underline{+\boxed{}} \quad \underline{+\boxed{}}$ Add $\boxed{}$ to both sides.

$5y = \boxed{}$

$\dfrac{5y}{\boxed{}} = \dfrac{25}{\boxed{}}$ Divide both sides by $\boxed{}$.

$y = \boxed{}$

LESSON 12-2 CONTINUED

Example 3

PROBLEM SOLVING APPLICATION

Troy has three times as many trading cards as Hillary. Subtracting 9 from the number of trading cards Troy has and then dividing by 6 gives the number of cards Sean has. If Sean has 24 trading cards, how many trading cards does Hillary own?

1. **Understand the Problem**
   Rewrite the question as a statement.

   - Find the number of trading cards that _____ has.

   List the important information:

   - Troy has _____ times as many trading cards as Hillary has.

   - Subtracting 9 from the number of trading cards that _____ has and then dividing by 6 gives the number of cards _____ has.

   - Sean has _____ trading cards.

2. **Make a Plan**

   Let c represent the number of trading cards Hillary has. Then $3c$ represents the number _____ has, and $\dfrac{3c - 9}{6}$ represents the number _____ has, which equals _____.

   Solve the equation $\dfrac{3c - 9}{6} = 24$ for c to find the number of cards _____ has.

LESSON 12-2 CONTINUED

3. Solve

$$\frac{3c-9}{6} = 24$$

$$(\quad)\frac{3c-9}{6} = (\quad)24 \qquad \text{Multiply both sides by } \underline{\quad}.$$

$$3c - 9 = \underline{\quad}$$

$$3c - 9 + \underline{\quad} = 144 + \underline{\quad} \qquad \text{Add } \underline{\quad} \text{ to both sides.}$$

$$3c = \underline{\quad}$$

$$\underline{\quad} = \underline{\quad} \qquad \text{Divide both sides by } \underline{\quad}.$$

$$c = \underline{\quad}$$

Hillary has _____ cards.

4. **Look Back**

If Hillary has _____ cards, then Troy has _____ cards. When you subtract 9 from 153, you get _____. And 144 divided by 6 is _____, which is the number of cards that _____ has. So the answer is correct.

Try This

1. Solve $14 - 8b + 12b = 62$.

2. Solve $4(y + 3) - 12 = 116$.

3. John is twice as old as Helen. Subtracting 4 from John's age and then dividing by 2 gives William's age. If William is 24, how old is Helen?

LESSON 12-3 Solving Equations with Variables on Both Sides

Lesson Objectives

Solve equations that have variables on both sides

Additional Examples

Example 1

Group the terms with variables on one side of the equal sign, and simplify.

A. $60 - 4y = 8y$

$60 - 4y +$ ☐ $= 8y +$ ☐   Add ☐ to both sides.

☐ $=$ ☐   Simplify.

B. $-5b + 72 = -2b$

$-5b +$ ☐ $+ 72 = -2b +$ ☐   Add ☐ to both sides.

☐ $=$ ☐   Simplify.

Example 2

Solve.

A. $7c = 2c + 55$

$7c -$ ☐ $= 2c -$ ☐ $+ 55$ Subtract ☐ from both sides.

☐ $=$ ☐   Simplify.

☐ $=$ ☐   Divide both sides by ☐.

$c =$ ☐

LESSON 12-3 CONTINUED

Example 3

Christine can buy a new snowboard for $136.50. She will still need to rent boots for $8.50 a day. She can rent a snowboard and boots for $18.25 a day. How many days would Christine need to rent both the snowboard and the boots to pay as much as she would if she buys the snowboard and rents only the boots for the season?

Let d represent the number of days.

$$18.25d = 136.5 + 8.5d$$

$18.25d - \boxed{} = 136.5 + 8.5d - \boxed{}$ Subtract ____ from both sides.

$\boxed{}\, d = \boxed{}$ Simplify.

$\dfrac{9.75d}{\boxed{}} = \dfrac{136.5}{\boxed{}}$ Divide both sides by ____.

$d = \boxed{}$

Christine would need to rent both the snowboard and the boots for ____ days to pay as much as she would have if she had bought the snowboard and rented only the boots.

Try This

1. Group the terms with variables on one side of the equal sign, and simplify.

   $-8b + 24 = -5b$

2. Solve.

   $54 - 3q = 6q + 9$

3. A local telephone company charges $40 per month for services plus a fee of $0.10 a minute for long distance calls. Another company charges $75.00 a month for unlimited service. How many minutes does it take for a person who subscribes to the first plan to pay as much as a person who subscribes to the unlimited plan?

LESSON 12-4: Inequalities

Lesson Objectives

Read and write inequalities and graph them on a number line

Vocabulary

inequality (p. 692)

algebraic inequality (p. 692)

solution set (p. 692)

compound inequality (p. 692)

Additional Examples

Example 1

Write an inequality for each situation.

A. There are at least 15 people in the waiting room.

"At least" means _____ than or _____ to.

B. The tram attendant will allow no more than 60 people on the tram.

"No more than" means _____ than or _____ to.

LESSON 12-4 CONTINUED

Example 2

Graph each inequality.

A. $n < 3$

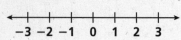

3 is not a solution, so draw an _____ circle at 3. Shade the line to the _____ of 3.

Example 3

Graph each compound inequality.

A. $m \leq -2$ or $m > 1$

First graph each _____ separately.

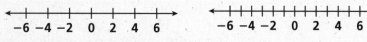

Then combine the graphs.

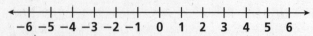

The solutions of $m \leq -2$ or $m > 1$ are the combined solutions of $m \leq -2$ and $m > 1$.

Try This

1. Write an inequality for the situation.

   There are at most 10 gallons of gas in the tank.

   _____

2. Graph the inequality.

   $p \leq 2$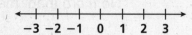

3. Graph the compound inequality.

   $5 > g \geq -3$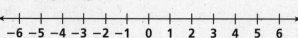

LESSON 12-5: Solving Inequalities by Adding or Subtracting

Lesson Objectives

Solve one-step inequalities by adding or subtracting

Additional Examples

Example 1

Solve. Then graph each solution set on a number line.

A. $n - 7 \leq 15$

$n - 7 \leq 15$

$+\underline{} +\underline{} $ Add ___ to both sides.

$n \leq$

←―|―――|―――|―――|―――|―――|―――|―――|→
  -14 -7 0 7 14 21 28 35

Draw a _____ circle at 22 then shade the line to the _____ of 22.

Example 2

Solve. Check each answer.

A. $d + 11 > 6$

$-\underline{} -\underline{} $ Subtract ___ from both sides.

$d >$

Check

$d + 11 > 6$

$\underline{} + 11 \stackrel{?}{>} 6 $ 0 is _____ than -5. Substitute 0 for d.

$11 \stackrel{?}{>} 6 \checkmark$

LESSON 12-5 CONTINUED

Example 3

Edgar's August profit of $137 was at least $20 higher than his July profit. What was July's profit?

August profit	was at least	$20 higher than	July's profit
$ ☐	≥	☐	+ p

$137 \geq 20 + p$

☐ − ☐   Subtract ☐ from both sides.

☐ ≥ p   Rewrite the inequality.

p ≤ ☐

Edgar's profit in July was at most $ ☐.

Try This

1. Solve. Then graph the solution set on a number line.

   $b - 14 \geq -8$

   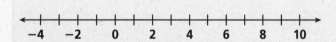

2. Solve. Check the answer.

   $a + 15 \leq 20$

3. Last year, the football team had at least 8 more takeaways than last year. Last year they had 10 takeaways. How many takeaways did they have this year?

LESSON 12-6: Solving Inequalities by Multiplying or Dividing

Lesson Objectives

Solve one-step inequalities by multiplying or dividing

Additional Examples

Example 1

Solve.

A. $\dfrac{c}{4} \leq -4$

$\dfrac{c}{4} \leq -4$

$()\dfrac{c}{4} \leq ()(-4)$ Multiply both sides by $\boxed{}$.

$c \leq \boxed{}$

Example 2

Solve. Check each answer.

A. $5a \geq 23$

$\boxed{} \geq \boxed{}$ Divide both sides by $\boxed{}$.

$a \geq \boxed{}$, or $\boxed{}$

Check

$5a \geq 23$

$5(\boxed{}) \stackrel{?}{\geq} 23$ 5 is $\boxed{}$ than $4\dfrac{3}{5}$. Substitute 5 for a.

$25 \stackrel{?}{\geq} 23$ ✓

LESSON 12-6 CONTINUED

Example 3

It cost Josh $85 to make candles for the craft fair. How many candles must he sell at $4.00 each to make a profit?

Since profit is the amount earned _____ the amount spent, Josh needs to earn _____ than $85.

Let c represent the number of candles that must be sold.

4c ☐ 85

$\frac{4c}{\Box}$ $\frac{85}{\Box}$ Divide both sides by ☐.

c ☐

Josh cannot sell 0.25 candle, so he needs to sell at least ☐ candles, or more than ☐ candles, to earn a profit.

Try This

1. Solve.

$\frac{r}{-3} > 0.9$

2. Solve. Check your answer.

$6b \geq 25$

3. It cost the class $15 to make cookies for the bake sale. How many cookies must they sell at 10¢ each to make a profit?

LESSON 12-7: Solving Two-Step Inequalities

Lesson Objectives
Solve simple two-step inequalities

Additional Examples

Example 1

Solve. Then graph each solution set on a number line.

A. $\dfrac{y}{2} - 6 > 1$

   +_____ +_____   Add _____ to both sides.

   $\dfrac{y}{2} >$ _____

   ( )$\dfrac{y}{2}$ > ( )7   Multiply both sides by _____.

   $y >$ _____

   ←—+—+—+—+—+—+—+—→
   −21 −14 −7 0 7 14 21

B. $5 \geq \dfrac{m}{-3} + 8$

   −_____ −_____   Subtract _____ from both sides.

   $-\dfrac{m}{3} \leq$ _____

   ( ) $-\dfrac{m}{3} \geq -3($)   Multiply both sides by _____, and _____ the inequality symbol

   $m \geq$ _____

   ←—+—+—+—+—+—+—→
   −3 0 3 6 9 12 15

LESSON 12-7 CONTINUED

Example 2

Sun-Li has $30 to spend at the carnival. Admission is $5, and each ride costs $2. What is the greatest number of rides she can ride?

Let r represent the number of rides Sun-Li can ride.

$5 + 2r \leq 30$

$\underline{-\boxed{} \quad -\boxed{}}$ Subtract $\boxed{}$ from both sides.

$2r \leq \boxed{}$

$\boxed{} \leq \boxed{}$ Divide both sides by $\boxed{}$.

$r \leq \boxed{}$, or $\boxed{}$

Sun-Li can ride only a whole number of rides, so the most she can ride is $\boxed{}$.

Try This

1. Solve. Then graph each solution set on a number line.

    $-9x + 4 \leq 31$

    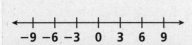

2. Brice has $30 to take his brother and his friends to the movies. If each ticket costs $4.00, and he must buy tickets for himself and his brother, what is the greatest number of friends he can invite?

Chapter Review

12-1 Solving Two-Step Equations

Solve. Check each answer.

1. $3x + 9 = 72$
2. $2q - 7 = 13$
3. $-4y + 11 = 75$

4. $38 = 7p - 18$
5. $\frac{z}{5} + 1 = 8$
6. $\frac{x}{-3} - 9 = 10$

7. A salesperson earned a paycheck for $2,750. The paycheck was a $500 bonus plus a flat rate for three seminars attended. What was the salesperson's rate of pay for each seminar?

12-2 Solving Multi-Step Equations

Solve.

8. $2y + 8 + 4y = 44$
9. $11c - 12 - 2c = 6$
10. $14 = -2x + 14 + x$
11. $3z + 4 - 6z = -20$
12. $3(x + 7) + 5 = 65$
13. $12 + 2(p - 5) = -6$

12-3 Solving Equations with Variables on Both Sides

Group the terms with variables on one side of the equal sign, and simplify.

14. $6x = 2x + 24$
15. $-4y + 10 = 6y$
16. $3c - 32 = -5c$

Solve.

17. $7x = 2x + 70$
18. $3p + 14 = -2p + 74$
19. $\frac{2}{5}x + 19 = \frac{1}{5}x + 21$
20. $-4 - y = 7y - 36$

CHAPTER 12 REVIEW *CONTINUED*

12-4 Inequalities

Write an inequality for each situation.

21. There are no more than 25 students in each class.

22. The height of that cliff is at least 500 feet.

Graph each compound inequality.

23. $x > 1$ or $x \leq -2$ 24. $-2 \leq x < 3$ 25. $4 < y \leq 7$

12-5 Solving Inequalities by Adding or Subtracting

Solve. Then graph each solution set on a number line.

26. $x + 1 < 10$ 27. $y - 3 \geq -2$ 28. $-12 < p - 8$

12-6 Solving Inequalities by Multiplying or Dividing

Solve.

29. $\dfrac{x}{6} < 2$ 30. $\dfrac{y}{-2} \geq 1$ 31. $-5y \geq 15$

32. JoAnne needs to raise $150. How many hours must she baby-sit at a rate of $4.50 per hour, in order to have enough money?

12-7 Solving Two-Step Inequalities

Solve. Then graph each solution set on a number line.

33. $3x + 3 \leq 30$ 34. $\dfrac{y}{5} + 7 > 19$ 35. $-15 \geq 3z + 6$

CHAPTER 12 Big Ideas

Answer these questions to summarize the important concepts from Chapter 12 in your own words.

1. Explain how to solve the equation $8x - 7 = 57$.

2. Explain how to solve the equation $4y + 9 - y = -3$.

3. Explain how to solve the equation $5z + 6 = -2z + 27$.

4. Explain how to solve the inequality $-8a \geq 32$.

5. Explain when to draw a closed circle or an open circle when graphing inequalities.

For more review of Chapter 12:

- Complete the Chapter 12 Study Guide and Review on pages 714–716 of your textbook.
- Complete the Ready to Go On quizzes on pages 690 and 708 of your textbook.